This volume presents the findings of a selection of pioneering research studies in which molecular techniques have been used to address key questions in biological anthropology, for example about the human genetic system, the geographical movements of human populations in the past, and primate evolution. Providing not only a timely overview of current research, this book also presents an insight into the potential significance of molecular biology in the decades to come, that will be of interest to all biological anthropologists as well as molecular biologists, human geneticists, palaeontologists and evolutionary biologists.

Cambridge Studies in Biological Anthropology 10

Molecular applications in biological anthropology

Cambridge Studies in Biological Anthropology

Series Editors

G. W. Lasker
Department of Anatomy, Wayne State University,
Detroit, Michigan, USA

C. G. N. Mascie-Taylor
Department of Biological Anthropology,
University of Cambridge

D. F. Roberts
Department of Human Genetics,
University of Newcastle upon Tyne

R. A. Foley
Department of Biological Anthropology,
University of Cambridge

Also in the series

G. W. Lasker   *Surnames and Genetic Structure*

C. G. N. Mascie-Taylor and G. W. Lasker (editors)   *Biological Aspects of Human Migration*

Barry Bogin   *Patterns of Human Growth*

Julius A. Kieser   *Human Adult Odontometrics – The Study of Variation in Adult Tooth Size*

J. E. Lindsay Carter and Barbara Honeyman Heath   *Somatotyping – Development and Applications*

Roy J. Shephard   *Body Composition in Biological Anthropology*

Ashley H. Robins   *Biological Perspectives on Human Pigmentation*

C. G. N. Mascie-Taylor and G. W. Lasker (editors)   *Application of Biological Anthropology to Human Affairs*

Alex F. Roche   *Growth Maturation, and Body Composition – The Fels Longitudinal Study 1929–1991*

Molecular applications in biological anthropology

EDITED BY

ERIC J. DEVOR

Department of Psychiatry,
University of Iowa College of Medicine, Iowa, USA

CAMBRIDGE
UNIVERSITY PRESS

CAMBRIDGE UNIVERSITY PRESS
Cambridge, New York, Melbourne, Madrid, Cape Town, Singapore, São Paulo

Cambridge University Press
The Edinburgh Building, Cambridge CB2 2RU, UK

Published in the United States of America by Cambridge University Press, New York

www.cambridge.org
Information on this title: www.cambridge.org/9780521391092

First published 1992
This digitally printed first paperback version 2005

A catalogue record for this publication is available from the British Library

Library of Congress Cataloguing in Publication data
Molecular applications in biological anthropology/edited by Eric J. Devor
    p.   cm. – (Cambridge studies in biological anthropology; 10)
    Includes index.
    ISBN 0-521-39109-1 (hard)
    1. Physical anthropology – Methodology.   2. Molecular biology –
    Technique.   I. Devor, Eric Jeffrey, 1949–   .   II. Series.
GN50.8.M65   1992
573–dc20   91-48061   CIP

ISBN-13 978-0-521-39109-2 hardback
ISBN-10 0-521-39109-1 hardback

ISBN-13 978-0-521-67552-9 paperback
ISBN-10 0-521-67552-9 paperback

To Dr Michael H. Crawford on whose shoulders I have been privileged to stand

Contents

Contributors

A. J. Boyce
Department of Biological Anthropology
University of Oxford, Oxford OX3 9DU, UK

J. B. Clegg
MCR Molecular Haematology Unit
Institute of Molecular Medicine, John Radcliffe Hospital, Oxford OX3 9DU, UK

E. J. Devor
Department of Psychiatry
University of Iowa College of Medicine, Iowa City, IA 42252, USA

J. Flint
MRC Molecular Haematology Unit
Institute of Molecular Medicine, John Radcliffe Hospital, Oxford OX3 9DU, UK

J. E. Hixson
Department of Genetics
Southwest Foundation for Biomedical Research, San Antonio, TX 78284, USA

G. A. Hoelzer
Department of Anthropology
Columbia University, New York, NY 10027, USA

R. L. Honeycutt
Department of Wildlife and Fisheries Sciences
Texas A & M University, College Station, TX 77843, USA

J. Marks
Departments of Anthropology and Biology
Yale University, New Haven, CT 06511, USA

D. J. Melnick
Department of Anthropology
Columbia University, New York, NY 10027, USA

J. Rogers
Department of Genetics
Southwest Foundation for Biomedical Research, San Antonio, TX 78228, USA

T. Turner
Department of Anthropology
University of Wisconsin–Milwaukee, Milwaukee, WI 53201, USA

M. L. Weiss
Program for Physical Anthropology
National Science Foundation, Washington, DC 20550, USA

Preface

The idea for producing this volume came during a symposium held in Kansas City, Missouri in the spring of 1989. At that time there were few molecular approaches to biological anthropology as the techniques of molecular biology were only just entering the field. Some pioneering studies were nevertheless under way, including those by most of the authors in this volume.

In the molecular revolution that is ongoing in biomedical research, disease and other genes are being mapped, cloned and sequenced at a pace unthinkable only a few years ago. There is an ongoing initiative to map and, eventually, sequence the entire human genome. Precise and sophisticated studies of genomic organization and gene regulation have been and are being carried out using the increasingly diverse armamentarium of molecular biology. Yet, the use of these tools to study questions in biological anthropology is only now becoming widespread. The unique perspective and straightforward focus of biological anthropology is also on the verge of a molecular revolution that will open avenues of research previously unavailable. The papers in this volume only hint at the possibilities to come.

To provide the necessary historical context, this collection opens with a short developmental history of one of the focal phenomena of current molecular research: the restriction fragment length polymorphism. In other areas of genetics there is a respectable time–depth with quantitative genetics going back to Galton in the last century, population genetics developing over the lifetime of Sewall Wright, and anthropological genetics starting up in earnest with the discoveries of Landsteiner early in the century. The molecular technology relevant to biological anthropology, on the other hand, is younger than most of those using it.

In the six chapters that follow that introduction, a wide range of applications of molecular techniques is presented. Rogers (Chapter 2) discusses the potential anthropological applications of DNA sequence variants. Hixson (Chapter 3) offers a specific application of molecular techniques in studying a non-human primate model of a human disease, atherosclerosis. From these studies a great deal is being learned about non-human primate genetics. Weiss and Turner (Chapter 4) focus on one particular class of DNA sequence variation – the VNTR/hypervariable – and show the many uses to which it may be put. In Chapter 5 Flint and colleagues turn their attention to the human globin gene clusters and show how molecular studies of these regions have opened up an enormous storehouse of information which can be applied to questions of both micro- *and* macro-evolution. Melnick and colleagues (Chapter 6) do much the same by presenting the 'other genome' – the mitochondrial DNA. Finally, Marks (Chapter 7) takes a closing look at the use of the entire genome as an entity for addressing broad evolutionary questions.

This volume has taken some time to assemble and finalize. The patience of the Cambridge University Press has been laudable. I wish to specifically acknowledge the efforts of Ms Rebecca Dill-Devor for producing edited copies of all of the chapters, Ms Nelly Mark for final typing of some of the chapters and for standardizing computer-ready materials, and Ms Roberta Rich for finalizing all of the artwork.

Eric J. Devor
Iowa City, Iowa

1 *Introduction: a brief history of the RFLP*

Introduction

Over the past few years, laboratory researchers have come to regard the use of molecular tools such as restriction endonucleases, charge modified nylon membranes, engineered cloning vectors, and the polymerase chain reaction (PCR) as almost commonplace and routine. In addition, the pace of development and implementation of both technology and technique is dizzying, e.g. restriction fragment length polymorphisms (RFLPs), DNA sequencing, chromosome mapping, PCR, cloning, transverse-field electrophoresis, hybrid cell lines, and transgenic animals. All this has, in turn, ushered in a molecular revolution so pervasive that no area of biology or medicine is untouched. Yet, the seminal discoveries which brought about this revolution are not even 25 years old.

In this volume, where many of the earliest uses of molecular techniques in addressing questions in biological anthropology are presented, it is useful to provide some historical context for the methods. Such an historical account must be selective, focusing upon the principle discoveries later utilized in the studies that are recounted here. Therefore, this chapter will concentrate upon the restriction fragment length polymorphism, RFLP. Clearly, this review will not be exhaustive but the timing of events and discoveries may come as a surprise to some. Finally, it is hoped that the basic information provided here will be useful in enhancing understanding of the chapters that follow.

The discovery of the restriction endonucleases

Restriction fragment length polymorphisms (RFLPs), the markers that have become so useful in all areas of molecular biology, are a subset of variable DNA sequences of base pairs that are detected by the use of DNA sequence-specific endodeoxyribonucleases. These enzymes recognize a specific nucleotide sequence and cleave both strands of the duplex molecule at or near the recognition site. These enzymes fall into three classes, based upon their molecular structure and their need for specific

1

cofactors. Class I endonucleases have a molecular weight around 300 000 Daltons, are composed of non-identical subunits, and require Mg^{2+}, ATP, and S-adenosyl-methionine (SAM) as cofactors for activity. Class II enzymes are much smaller, with molecular weights in the range of 20 000 to 100 000 Daltons. They have identical subunits, and require only Mg^{2+} as a cofactor (Nathans and Smith, 1975). The Class III enzyme is the rarest of the three types. It is a large, heterodimer with a molecular weight around 200 000 Daltons and it differs from the Class I enzyme by not needing SAM as a cofactor for endonuclease activity. It does, however, require ATP as well as Mg^{2+}.

In 1968 nothing about these enzymes was known because their existence was unknown. In that year, Matthew Meselson and Robert Yuan reported the discovery of an enzyme in the bacterium *Escherichia coli*, strain K-12, that appeared to be able to recognize and digest foreign DNA, i.e. bacteriophage grown in other *E. coli* strains. This enzyme, they concluded, could be the agent responsible for the phenomenon of 'restriction', the ability of many strains of *E. coli* to recognize and degrade DNA from foreign strains (Meselson and Yuan, 1968). They named this enzyme a **restriction endonuclease** and determined that such enzymes would recognize and degrade any DNA that did not contain a specific pattern of methylation of its nucleotides (Meselson and Yuan, 1968: 1110). Their *E. coli* enzyme was found to require the presence of magnesium ions (Mg^{2+}), ATP, and SAM to be active. Thus, appropriately, the first restriction enzyme to be identified was a Class I enzyme.

The report of Meselson and Yuan was quickly followed by two papers, by Smith and Wilcox (1970) and Kelly and Smith (1970), describing a similar enzyme in the bacterium *Haemophilus influenzae*, strain Rd. Like the *E. coli* enzyme, the *H. influenzae* endonuclease was inactive in the presence of native DNA but did recognize and digest foreign DNAs. Unlike the *E. coli* enzyme, however, the Smith and Wilcox discovery was fully active when only Mg^{2+} was present. The cleavages of both enzymes were limited in number and consistent, suggesting a specific DNA recognition sequence and binding of the enzyme at that site prior to cleavage (Smith and Wilcox, 1970: 390). In the companion paper, Kelly and Smith offered evidence that the recognition site of their endonuclease was a run of six specific nucleotides,

$$5' \ldots \text{G T Py} \mid \text{Pu A C} \ldots 3'$$
$$3' \ldots \text{C A Pu} \mid \text{Py T G} \ldots 5'$$

where Py refers to either pyrimidine and Pu to either purine. The rotational symmetry of this sequence did not escape notice: 'It is unlikely that the symmetry of this sequence is fortuitous, since the number of possible asymmetric sequences of this type is about 30 times the number of possible symmetrical sequences . . .' (Kelly and Smith, 1970: 407). Thus, they concluded that symmetry in the recognition sequence was a basic feature of the action of the enzyme.

By 1970, the general outlines of endonuclease action were known. These seminal papers opened the floodgates and, by 1975, Daniel Nathans and Hamilton Smith were able to review the state of knowledge and present a consistent, standard nomenclature for the rapidly growing number of known restriction endonucleases. The name of each enzyme would convey the genus and species of the bacterium from which the enzyme was isolated, the strain number of that species, and the order in series in which the enzyme was found. Thus, *Eco RI* refers to *E. coli*, strain R, first enzyme (Table 1.1). As the list of known restriction enzymes grew, it was found that more than one enzyme could recognize the same run of nucleotides. To these enzymes R. J. Roberts gave the name **isoschizomers** (Nathans and Smith, 1975).

Knowledge of existence of restriction enzymes and the quick identification of unusual and interesting aspects of their recognition sites set off a massive search among the then known bacterial stocks such that, by 1982, a list of 357 identified restriction enzymes recognizing more than 90 different DNA sequences could be made (Linn and Roberts, 1982). Recognition sequences of from four to eight bases are known. Additional diversity was found among the isoschizomers. For example, the enzymes *Sma I* and *Xma I* both recognize the six base sequence CCCGGG but give different fragments, with *Sma I* cutting in between the third and fourth bases (CCC/GGG) and *Xma I* cutting in between the first and second (C/CCGGG). Similarly, the isoschizomer pair *Hha I* and *Hin PI* both recognize the four base sequence GCGC but cut differently: *Hha I* (GCG/C) and *Hin PI* (G/CGC). Further differences were found with respect to sensitivity to methylation. Both *Mbo I* and *Sau 3A* recognize the four-base sequence GATC and will cut in the same place (GA/TC). However, when the sequence is modified by methylation of the adenosine, GA*TC, *Mbo I* will not cut and *Sau 3A* will. When the cytosine is methylated, GATC*, then *Mbo I* cuts and *Sau 3A* fails. A very useful isoschizomeric pair in this regard is *Hpa II* and *Msp I*. These enzymes recognize the four-base sequence CCGG. When this sequence is modified as CC*GG, *Hpa II* will not cut while *Msp I* does. This pair of enzymes has proved to be extremely valuable in the identification of CpG, or methylation, 'islands' that lie near coding genes (cf. Bird, 1986, 1989).

Table 1.1. *A compilation of seventy-five restriction enzymes giving the microorganism from which each was isolated, the recognition sequence, mean predicted fragment sizes, and known isoschizomers*

Restriction enzyme	Microorganism	Recognition sequence[a]	S[b]	Class[c,d]	Isoschizomers[c]
Aat II	Acetobacter aceti	GACGT/C	38 600	VIII	
Acc III	Acinetobacter calcoaceticus	T/CCGGA	18 400	VII	Bsp EI, Bsp MII, Mro I
Afl II	Anabaena flos-aquae	C/TTAAG	2700	IV	Bfr I
Alu I	Arthrobacter luteus	AG/CT	180	I	
Apa I	Acetobacter pasteurianus	GGGCC/C	6350	V	Bsp 120 I
Asu I	Anabaena subcylindrica	G/GNCC	400	I	Sau 96I, Cfr 131, Nap IV
Ava I	Anabaena variabilis	C/PvCGPuG	11 700	VII	Aqu I, Nsp III
Ava II	Anabaena variabilis	G/G(A/T)CC	820	II	Afl I, Sin I, Eco 47I
Ava III	Anabaena variabilis	ATGCAT	2300	III	Nsi I, Eco T22I
Bal I	Brevibacterium albidum	TGG/CCA	2750	IV	Msc I
Bam HI	Bacillus amyloliquefaciens H	G/GATCC	5400	V	Bst I
Bbv I	Bacillus brevis	GCAGC(N)8/12	1230	II	
Bcl I	Bacillus caldolyticus	T/GATCA	2300	III	
Bgl I	Bacillus globigii	GCC(N)4/NGGC	8900	VI	
Bgl II	Bacillus globigii	A/GATCT	2250	III	
Bss HII	Bacillus stearothermophilis H3	G/CGCGC	189 000	X	Bse PI
Bst EII	Bacillus stearothermophilis ET	G/GTNACC	6950	V	Eco 91I
Bst NI	Bacillus stearothermophilis	CC/(AT)GG	480	I	Apy I, Mva I, Eco RII
Cla I	Caryophanon latum L	AT/CGAT	15 800	VII	Ban III, Bsc I, Asp 707, Bsu 15I
Dde I	Desulfovibaio desulfuricans	C/TNAG	210	I	
Dra I	Deinococcus radiophilus	TTT/AAA	1420	II	Aha III
Eco RI	Escherichia coli RY13	G/AATTC	3200	IV	Rsr I
Eco RV	Escherichia coli J62	GAT/ATC	5500	V	Eco 32I
Fnu DII	Fusobacterium nucleatum D	CG/CG	9000	VI	Acc II, Bst UI, Tha I, Mvn I
Fnu 4HI	Fusobacterium nucleatum 4H	GC/NGC	535	I	
Hae II	Haemophilus aegyptius	PuGCGC/Pv	11 100	VII	

Enzyme	Organism	Recognition sequence	Group	Number	Isoschizomers
Hae III	Haemophilus aegyptius	GG/GC	I	380	Fnu DI, Bsu RI, Pal I
Hha I	Haemophilus haemolyticus	GCG/C	III	2100	Hin PI, Cfo I, Hin 6I
Hind III	Haemophilus influenzae Rd	A/AGCTT	III	1750	
Hinf I	Haemophilus influenzae Rf	G/ANTC	I	300	
Hpa I	Haemophilus parainfluenzae	GTT/AAC	V	5050	
Hpa II	Haemophilus parainfluenzae	C/CGG	III	1650	Hap II, Msp I
Hph I	Haemophilus parahaemolyticus	GGTGA(N)8/7	II	1090	
Kpn I	Klebsiella pneumoniae OK8	GGTAC/C	VI	8500	Asp 718, Sth I
Mae I	Methanococcus aeolicus	C/TAG	I	275	Rma I
Mae II	Methanococcus aeolicus	A/CGT	II	1460	
Mbo I	Moraxella bovis	/GATC	I	325	Nde II, Dpn II, Sau 3AI
Mla I	Mastigocladus laminosus	TT/CGAA	VII	10900	
Mlu I	Micrococcus luteus	A/CGCGT	X	132000	
Mnl I	Moraxella nonliquefaciens	CCTC(N)7/7	I	295	
Msp I	Moraxella species	C/CGG	III	1650	Hpa II, Hap II
Mst I	Microcoleus species	TGC/GCA	VII	14900	Fsp I, Aos I, Fdi II
Nae I	Nocardia aerocolonigenes	GCC/GGC	VIII	34600	
Nar I	Nocardia argentinensis	GG/CGCC	VIII	34600	Nun II, Bbe I
Nci I	Neisseria cinerea	CC/(G/C)GG	IV	3300	Aha I, Bcn I, Cou II
Nco I	Nocardia corallina	C/CATCG	IV	3400	
Nde I	Neisseria denitrificans	CA/TATG	IV	3500	
Nla III	Neisseria lactamica	CATG	I	210	
Not I	Nocardia otitidis-caviarum	GC/GGCCGC	XI	3 000 000	
Nru I	Nocardia rubra	TCG/CGA	X	100 500	
Pac I	Pseudomonas alcaligenes	TTAAT/TAA	VII	20 900	
Pst I	Providencia stuartii 164	CTGCA/G	IV	3050	
Pvu I	Proteus vulgaris	CGAT/CG	X	126 500	Xor II, Xml I, Bsp CI
Pvu II	Proteus vulgaris	CAG/CTG	IV	3050	
Rsa I	Rhodopseudomonas sphaeroides	GT/AC	I	515	Csp 6I
Sac I	Streptomyces achromogenes	GAGCT/C	V	4800	Sst I, Ecl 136I
Sac II	Streptomyces achromogenes	CCGC/GG	X	148 500	Sst II, Ksp I, Cfr 42I
Sal I	Streptomyces albus G	G/TCGAC	VIII	38 600	Sal GI
Sau 3A	Staphylococcus aureus 3A	/GATC	I	325	Dpn II, Mbo I, Nde II

Table 1.1. (*cont.*)

Restriction enzyme	Microorganism	Recognition sequence[a]	S[b]	Class[c,d]	Isoschizomers[e]
Sca I	Streptococcus caespitosus	AGT/ACT	4050	IV	
Scr FI	Streptococcus cremoris F	CCNGG	420	I	Dsa V, Sso II
Sfi I	Streptomyces fimbriatus	GGCC(N)$_4$/NGGCC	149 000	X	
Sma I	Serratia marcescens S	CCC/GGG	27 200	VIII	Xma I, Xcy I, Cfr 9I
Sna I	Sphaerotilus natans C	GTATAC	8750	VI	Xca I
Sna BI	Sphaerotilus natans	TAC/GTA	30 000	VIII	Eco 105I
Sph I	Streptomyces phaeochromogenes	GCATG/C	4350	IV	Bbu I, Pae I
Ssp I	Sphaerotilus species	AAT/ATT	1820	II	
Sst I	Streptomyces stanford	GAGCT/C	4800	V	Sac I, Ecl 136I
Stu I	Streptomyces tubercidicus	AGG/CCT	3000	IV	Aat I, Eco 147I
Taq I	Thermus aquaticus YTI	T/CGA	1110	II	Tth HB8I
Tth 111 II	Thermus thermophilus 111	CAAPuCA(N)11/9	1070	II	Asp I
Xba I	Xanthomonas badrii	T/CTAGA	3050	IV	
Xho I	Xanthomonas holcicola	C/TCGAG	20 600	VII	Pae R7I, Sex I
Xma I	Xanthomonas malvacearum	C/CCGGG	27 200	VIII	Sma I, Xcy I, Cfr 9 I
Xma III	Xanthomonas malvacearum	C/GGCCG	149 000	X	Eag I, Ecl XI, Eco 52I

[a]The single code letter 'N' refers to any of the four bases A, C, G, T; Pu refers to the purines C or T, Py refers to the pyrimidines G or A.

[b]S is the predicted average restriction fragment size based upon the recognition sequence (Drmanac et al., 1988).

[c]Classes from Drmanac et al. (1988) refer to groups based on S. Class I groups from 100 to 750 bp, Class II 751–1500 bp, Class III 1501–2500 bp, Class IV 2501–4500 bp, Class V 4501–7500 bp, Class VI 7501–10 000 bp, Class VII 10 001–25 000 bp, Class VIII 25 001–50 000 bp, Class X 100 001+ bp. No Class IX enzymes are known and Class XI contains only the enzyme Not I.

[d]As a general rule, the class to which an enzyme belongs will influence decisions about the agarose gels used to size resolve restriction fragments and the subsequent Southern transfer of those fragments. Enzymes in Class I and II usually require 1.0–2.0% agarose gels, Class III–V usually resolve better on 0.7–0.8% gels and often the gels need to be acid nicked (0.25 M HCl) prior to transfer, and Class VI and above are best resolved by the non-standard transverse field gel electrophoresis techniques (see Lai et al., 1989).

[e]The lists of isoschizomers contain both enzymes that cut the recognition sequence in the same place and enzymes which recognize the same sequence but cut in a different place in that sequence.

As is seen in Table 1.1, most restriction endonucleases recognize GC-rich sequences. However, some AT-only sequences are known. These include enzymes such as *Dra I* (TTTAAA), *Ssp I* (AATATT), and *Pac I* (TTAATTAA). Further, most restriction enzymes cleave DNA within the recognition sequence itself but several do not. For example, the enzyme *Mnl I* recognizes the non-palindromic four base sequence CCTC but it cuts seven base pairs downstream (CCTC 7/7). Other such enzymes include *Bbv I* (GCAGC 8/12) and *Hga I* (GACGC 5/10). Other variations on the basic theme include the 'degenerate' recognition sequences. These are sequences that permit more than one base to occupy a particular position. For example, the enzyme *Acc I* recognizes the sequence GT(A/C)(T/G)AC, meaning that both GTATAC and GTCGAC will be cut by this enzyme. The five-base cutter *Hinf I* recognizes the sequence GANTC, meaning that the middle base in the sequence can be any of the four nucleotides. This last form of degenerate sequence can become quite large as seen in the recognition sites of *Hgi EII* (ACC(N)$_6$GGT) and *Sfi I* (GGCC(N)$_4$GGCC), where any run of four or six bases is allowed.

The nature of the sequence recognized by a restriction enzyme will influence the frequency at which it will cleave a target DNA. Bishop *et al.* (1983) derived a model for the average expected fragment length, in base pairs, produced by a number of restriction enzymes as a function of their recognition sequences. Given the observations of DNA-based dimer frequencies reported by Arthur Kornberg and colleagues (Swartz *et al.*, 1962) that showed extreme non-randomness in eukaryotic DNAs, Bishop *et al.* created a matrix of dimeric transition frequencies. For example, given the base adenosine in the first position, the probability that it is followed by another adenosine is 0.32. Given the base cytosine in the first position, the probability that it is followed by guanosine is only 0.05 (Bishop *et al.*, 1983: 799). Using the full set of dimer transition frequencies, the recognition sequence of the enzyme *Sau 3A* (GATC) should occur quite often. Simplifying, the transition probabilities G → A = 0.30, A → T = 0.27, T → C = 0.19 are reasonably common so that, in an average eukaryotic genome, *Sau 3A* cuts every 318 bp. At the other end of the scale, the rare C → G transition makes the four-base recognition sequences of *Taq I* (ACGT) and *Msp I* (CCGG) less common. Consequently, these enzymes cut every 1179 bp and every 1747 bp respectively. Longer sequences such as that recognized by *Xma I* (CGGCGG) yield many fewer sites and longer distances between, as here at 166 517 bp.

Enzymes having degenerate recognition sequences have an inherent flexibility that translates into smaller mean fragment sizes than any single sequence could give. For example, the enzyme *Hinf I* (GANTC) has four possible recognition sites: GAATC, GACTC, GAGTC, and GATTC. Bishop *et al.* (1983: 802) show that the mean fragment sizes of these four sequences in the average eukaryotic genome are 994, 1364, 1556 and 964 bp, respectively. However, since these sites can occur in any order, the $A \rightarrow N$ transition is 1.00 and is replaced by $A \rightarrow N \rightarrow T = 0.27$, giving a mean per enzyme fragment length of 292 bp. A similar analysis of more than 80 restriction endonucleases was carried out by Drmanac *et al.* (1988). The results of their analyses indicated rough groupings of enzymes based upon their recognition sequences. Group I contained all enzymes whose recognition sequences would be expected to occur every 500 bp or less. Each subsequent group was demarcated in a reasonably clear-cut manner (Table 1.1). Such information is useful in a variety of experiments on eukaryotic DNA.

The discovery of the RFLP

While restriction endonucleases are of great interest in their own right and much can be learned from them, it is as reagents in DNA studies that makes them universally useful. One of the first applications of restriction enzymes as reagents was to produce a restriction site map of the rabbit β-globin gene (Jeffreys and Flavell, 1977). The map was constructed by carrying out a series of single and multiple restriction enzyme digests of rabbit genomic DNA, followed by electrophoretic separation of the resulting restriction fragments in agarose gels. These size-separated fragments were then transferred to a nitrocellulose support via the method of Southern (1975). A radiolabelled plasmid containing rabbit β-globin mRNA was used as the probe. Analysis of the hybridizing fragments revealed the number and order of the restriction enzyme recognition sites and the distances between them. When this technique was applied to the human β-globin gene region on chromosome 11, a number of restriction enzyme cleavage sites were found in a 300 bp sequence. Most importantly, three of these sites, one for the enzyme *Pst I* and two for the enzyme *Hind III*, were observed to be polymorphic between individuals (Jeffreys, 1979). The nature of the polymorphism was seen in the lengths of the restriction fragments produced. Thus, the name given to this new form of variation was restriction fragment length polymorphism, or RFLP.

Reflecting on this new source of variation, Jeffreys notes (1979: 1), 'There is, however, almost no information on how much variation exists at the level of DNA sequences in man, and on what types of DNA sequence variants might occur in human populations.' The detection of three RFLPs in so small a region of the β-globin gene suggested that the amount of heritable DNA sequence variation could be enormous. Thus, 'DNA sequence variants would serve as useful new markers in examining the population structure and origins of human races . . .' (Jeffreys, 1979: 8). Jeffreys's RFLPs were confirmed and expanded upon by Tuan *et al.* (1979) but another RFLP, for the enzyme *Hpa I*, also in the human β-globin region, actually has the distinction of being the first human RFLP identified (Kan and Dozy, 1978). Numerous heritable reproducible sequence variants in human DNA have been found and named.

At the time the first human RFLPs were being described, a revolutionary idea emerged. If these new DNA sequence variants occurred throughout the human genome and in sufficient quantity, they could be used to map the entire genome (Botstein *et al.*, 1980). Up to that time the precious few RFLPs known were all associated with structural genes and were not very informative except when taken in haplotype groups. Not daunted by the fact that no evidence existed for the presence in the human genome of RFLPs detected by random, anonymous DNA segments, Botstein *et al.* suggested that 'The advent of recombinant DNA technology has suggested a theoretically possible way to define an arbitrarily large number of arbitrarily polymorphic marker loci' (Botstein *et al.*, 1980: 315).

The concept of informativeness also found a new meaning in the seminal paper by Botstein *et al.* In that paper the value known as the polymorphism information content, or PIC value, was introduced. The PIC value is meant to convey, in a single number whose range is zero to one, the usefulness of any polymorphism for the purpose of mapping. PIC is defined as

$$1 - \left(\sum_{i=1}^{n} p_i^2\right) - \sum_{i=1}^{n-1} \sum_{j=i+1}^{n} 2p_i^2 p_j^2 \tag{1}$$

where p_i and p_j represent the frequencies of the ith and j^{th} alleles of any RFLP. This value has become the gold standard by which utility of all polymorphisms is judged. For any two allele RFLP the maximum PIC is 0.38 which is obtained when $p_i = p_j = 0.5$. Such a value, by the standards issued by Botstein *et al.* and subsequently borne out by experience, is only moderately informative for mapping and linkage. Values below 0.3 are regarded as relatively uninformative except in very large samples while values above 0.5 are what everyone wants because this indicates a highly informative multiple allele system.

In 1980, genetic systems with PIC values above 0.4 were very rare. These included the six-allele Rh blood group system (PIC = 0.58) and the multiple allele HLA systems (PIC → 0.98). Yet Botstein *et al.* believed that arbitrary RFLPs detected by anonymous DNA segments must exist in the human genome as it made little sense that such variants would be restricted only to small regions near structural genes. Their faith was rewarded within the year when a series of arbitrary clones from an *Eco RI* human genomic library were screened. One clone, labelled AW101, detected a highly informative RFLP (Wyman and White, 1980). This clone is perhaps the most historically important piece of DNA ever found as it simultaneously proved two of the critical assertions of the Botstein *et al.* (1980) paper. First, it was not associated with any known structural gene as it was picked out of a total genomic library. Secondly, it was very polymorphic with at least eight alleles observed immediately (Wyman and White, 1980). Thus, it was true that highly informative, anonymous DNA markers could be found.

From this promising beginning came a flood of RFLPs. Barker *et al.* (1984) screened 16 restriction enzymes with 31 cloned human DNA probes. Nine of these probes detected RFLPs in panels of unrelated individuals. Moreover, even at this very early stage a pattern began to emerge. Among the ten RFLPs discussed by Barker *et al.* using their nine probes, only three of the 16 restriction enzymes screened were represented. These were *Msp I* (CCGG) six times, *Taq I* (TCGA) three times, and *Bgl II* (AGATCT) once. The one feature shared by any of these enzymes was that both *Msp I* and *Taq I* contained the rare CpG dimer in their recognition sequence and they were the only enzymes in their screening panel to have it. *Bgl II* was becoming known as a good source of RFLPs but for quite different reasons (Pearson *et al.*, 1982). Thus, Barker *et al.* (1984: 135–6) suggested that restriction enzymes containing CpG dimers in their recognition sequences would be excellent candidates for RFLP screening. This suggestion was based upon a model of point mutation in which the cytosine of a CpG dinucleotide sequence becomes preferentially methylated forming 5-methylcytosine. This compound is known to undergo spontaneous denaturation *in vivo* leaving a thymidine base as the result (Razin and Riggs, 1980). Thus, CpG-containing restriction enzyme recognition sequences would be destroyed by point mutation at a much higher rate than other sequences, resulting in a preference for these enzymes to reveal RFLPs. Subsequent empirical observations by Youssoufian *et al.* (1986) and others verified this model.

The demonstration that restriction enzymes containing CpG dimers in their recognition sequence were hotspots for RFLP detection was of only limited utility, however. The rarity of the CpG dimer in eukaryotic DNA meant that restriction enzymes other than *Taq I* and *Msp I* having it in their recognition sequences would cut relatively infrequently. This would result in restriction fragments of sizes beyond the ability of conventional agarose gel electrophoresis to resolve. For example, following Drmanac *et al.* (1988), *Taq I* yields an average predicted fragment size of 1110 bp and *Msp I* yields an average predicted fragment size of 1650 bp. The next set of restriction enzymes with CpG dimers in their recognition sequences includes *Ava I*, (CPyCGPuG) *Cla I*, (ATCGAT) and *Xho I* (CTCGAG) whose mean predicted fragment sizes range from 11 700 bp to 20 600 bp. Beyond this are enzymes producing average fragment sizes of 30 000 bp (*Sna BI*, TACGTA), 38 600 bp (*Sal I*, GTCGAC), 126 500 bp (*Pvu I*, CGATCG), and up to *Not I* (GCGGCCGC) at 3×10^6 bp. Clearly, to be useful in conventional RFLP screening, non-CpG containing sequences would have to also be polymorphic.

From the work of Barker *et al.* (1984) and others it appeared that some enzymes were better than others in revealing RFLPs. The list of 'good' restriction enzymes did include some which did not have any CpG dinucleotides in their recognition sequences. But were there any rules or general principles governing restriction enzyme efficiency? Wijsman (1984) attempted to model the relative efficiencies of restriction enzymes in detecting DNA sequence variation. One feature to emerge at once was that the size and specific base order of the recognition sequence *did* have an influence on efficiency. Four-base cutters would cleave DNA more often than would six-base cutters, thereby creating more opportunities for detecting sequence variants. Wijsman showed two other critical factors as well. One was the size of the DNA probe being used: a 2 kb probe was better than a 12 kb probe since small variations would tend to be obscured when using a larger probe. The second factor was more subtle: the size of the smallest fragment that could be detected had a large effect. If fragments down to 200 bp could be resolved in the hybridization, the efficiency of the vast majority of restriction enzymes in revealing RFLPs would be increased compared to a situation where the smallest detectable fragment size was 500 bp. Interestingly, both of these factors had a greater effect upon the more efficient enzymes than upon the less efficient based on consideration of recognition sequence alone (Wijsman, 1984: 9216–17).

Taking data from RFLP reports published in the journal *Nucleic Acids Research* in 1985 and 1986, Devor (1988) used Wijsman's model to update the relative efficiency of restriction enzymes. As predicted *Taq I*, *Bgl II*, and *Hae III* did very well while *Kpn I* was significantly disappointing. Thus, Wijsmans' model had held up. In addition, two other, general features of RFLP searches came to light. Helentjaris and Gesteland (1983) showed that cDNA probes were less efficient than were genomic clones in detecting RFLPs. This was, of course, owing to the fact that cDNAs are not intact stretches of DNA while genomic clones are. This was especially true if the cDNA was small. Additionally, for both types of probe, the X-chromosome appeared to be less variable than any of the autosomes (Hofker *et al.*, 1986). Thus, in the early scramble to find and map RFLPs, the optimal strategy appeared to be to use large, genomic clones to screen frequent-cutting enzyme digests (Feder *et al.*, 1985).

By the time of Ninth International Conference on Human Gene Mapping (HGM9, 1987) more than 2000 anonymous DNA segments were known of which nearly half were polymorphic. These figures represented more than 80% of the polymorphic DNA segments that had been described in the seven years since the Botstein *et al.* (1980) paper. At the Tenth International Conference on Human Gene Mapping, only two years later (HGM10, 1989), these numbers had grown to more than 3000 with the number of polymorphic sequences nearly doubled. While this would seem to have been a fulfilment of the ideas laid out in that paper, a new and less appealing fact about these sequence variants was becoming clear. The vast majority of RFLPs had only two alleles and few had PIC values near the 0.38 maximum such polymorphisms could have. Lange and Boehnke (1982) had followed on the original Botstein *et al.* ideas with an estimate that the human genome could be spanned by a linkage map composed of 165 evenly spaced RFLP markers. Their estimate tacitly assumed that these markers would be highly informative. Bishop *et al.* (1983) approached the problem of coverage somewhat more systematically. Given a genome estimated to be 3300 cM (centiMorgans, a map unit roughly equal to 1 million bp), Bishop *et al.* showed that a map of that genome to 20 cM resolution would require 766.3 DNA markers. Reasonable coverage could be achieved with 400 markers. Unfortunately, even by HGM9, many of the RFLPs being reported were simply not very useful for mapping. Clearly, what was needed were more informative markers.

Minisatellites and VNTRs

The strategy used by Wyman and White (1980) and by Barker *et al.* (1984) to screen anonymous DNA clones from libraries for RFLPs was paying off handsomely by 1985 in the form of hundreds of mapped polymorphisms. The vast majority of these were being detected by DNA sequences that were unique in the human genome, that is, the polymorphic probes being developed were single-copy probes capable of detecting two types of variation. One was the traditional RFLP caused by the gain or loss of a restriction endonuclease recognition site and the other was an RFLP caused by the insertion or deletion of genetic material between two invariant restriction enzyme recognition sites. The human genome is generally regarded as being 3.2×10^9 bp in length. Jeffreys (1979) and Cooper *et al.* (1985) suggested that the total number of sequence variants within the human genome could be between 10^7 and 3×10^7. The conventional RFLPs being detected by single-copy stretches of genomic DNA may account for only 1% of this variation (Cooper *et al.*, 1985; Wallace *et al.*, 1986). The insertion–deletion type of RFLPs are less common by at least an order of magnitude than the conventional restriction site change RFLP (Wijsman, 1984). Thus, a vast store of heritable genomic variation exists that the RFLP probes being developed in 1985 could not detect.

During their study of the human myoglobin gene on chromosome 22, Jeffreys and colleagues found an unusual stretch of repetitive DNA in one of the introns. A 33 bp sequence was seen to repeat four times with perfect fidelity. This tandem repeat was flanked by a perfect 19 bp interspersed repeat. This repetitive DNA sequence had the even more unusual property of hybridizing to a large number of restriction fragments of a genomic digest rather than to just one (Weller *et al.*, 1984). More important, the numbers and sizes of the fragments to which this sequence hybridized were highly variable among unrelated individuals but appeared to be Mendelian when followed in families. This was a new type of DNA probe to which Jeffreys *et al.* (1985a) gave the name 'minisatellite' to refer to the relatively low copy number of the tandem repeat. Jeffreys used the core sequence of the myoglobin repeat to generate a pure repeat containing probe called pAV33.7 which was, in turn, used to screen a human genomic library. Forty positive clones were identified and eight were chosen for RFLP screening. All eight clones were found to be extremely polymorphic. The new type of polymorphism, named hypervariable by Jeffreys, was soon shown to be due not to

A Conventional RFLP

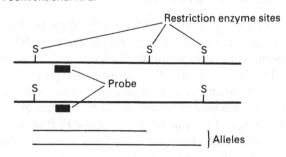

B VNTR polymorphism

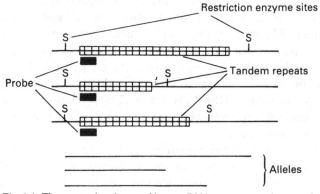

Fig. 1.1. The two major classes of known DNA sequence variant are shown. The conventional RFLP (*A*) is created when a restriction enzyme recognition site, S, is either gained or lost by point mutation. The variable number of tandem repeat, or VNTR, polymorphism (*B*) is created when varying numbers of copies of a repeated core sequence are present between two restriction enzyme sites. Since the probes used to detect VNTRs are specific for the repeat-containing core sequence, the numbers of alleles possible are far greater than for the conventional polymorphism which must rely upon relatively rarer mutational events (cf. Nakamura *et al.*, 1987).

insertion–deletion or the gain or loss of restriction enzyme recognition sequences but, rather, to variable numbers of copies of the core sequence repeated between restriction sites (Fig. 1.1). This type of polymorphism in repetitive DNA could prove enormously useful since repetitive DNA sequences make up as much as one half of the human genome.

The early promise of the hypervariable minisatellite probes was to provide a large number of simultaneously detectable loci which could lead to wholesale mapping of disease loci (Jeffreys *et al.*, 1986). While these probes have become tremendously valuable in other ways, this early hope rapidly faded. The primary reasons for this disappointment were that (1) unlike the conventional, single-copy RFLPs, a result from one family for any minisatellite could not be directly compared with results from any other family for the same probe and (2) most true minisatellites were found to have very high mutation rates (Jeffreys *et al.*, 1986, 1988). Thus, the hypervariation that made these probes so attractive at the outset led to their ultimate demise as linkage markers. It was not long, however, before these problems were overcome by the development of yet another new type of DNA probe.

The dual problems of informativeness and non-additivity suffered by single-copy and hypervariable DNA probes respectively would be solved if a way could be found to generate single-copy, or at least single-locus, hypervariables. The Wyman and White probe, pAW101, appeared to be one such case and Jeffreys's group showed that more probes of that type could be derived from tandem repeat sequences. In their first attempt a specific hybridizing fragment from a hypervariable minisatellite was isolated in a gel and purified for cloning. This clone, called pλg3, was found to hybridize to a single locus in the human genome and revealed an astounding 77 alleles among the first 79 unrelated individuals screened (Wong *et al.*, 1986). Following on the understanding gained from this success, Wong *et al.* (1987) derived several more single locus hypervariable clones by screening a human size-selected genomic library with their minisatellite probes. These new clones all proved to be as polymorphic as pλg3 and, most importantly, they showed that sequences of this type might be very abundant in the human genome. Knowlton *et al.* (1986) screened more than 1600 randomly selected recombinant clones from a human genomic library and found a dozen that were hyper-polymorphic. If this library could be considered representative, then more than 1500 hyper-polymorphic regions would be found in the human genome (Wong *et al.*, 1987: 270).

Armed with this information a group headed by Ray White set about to clone large numbers of hypervariable probes from human genomic libraries. The strategy selected was to synthesize oligonucleotides of up to 20 bases reflecting the core sequences of known hypervariables including two myoglobin regions and one each from zeta-globin and insulin. These oligonucleotides were then used to screen the genomic libraries under reduced stringency conditions which would permit mismatches in the hybridizing sequences. Under these conditions more than 300 clones were tested for polymorphism. Seventy-seven revealed polymorphisms of from two to more than 20 alleles (Nakamura *et al.*, 1987). Most of these clones belonged to the same class of DNA probe as those derived by Wong *et al.* (1987). The name given to this class of probe and polymorphism was variable number of tandem repeat, or VNTR (Nakamura *et al.*, 1987: 1617). Today, several hundred VNTR probes have been characterized and mapped and more are being developed including a subclass based on PCR (Weber and May, 1989). The rapid growth of VNTRs over the past few years attests to the optimistic estimates offered by Wong *et al.* (1987).

Summary

The first restriction enzymes were discovered in bacteria in 1968 and the first human RFLP was reported ten years later. Over the following decade the pace of discovery increased dramatically. By the Tenth International Congress on Human Gene Mapping (HGM10, 1989) a total of 1886 RFLPs had been reported (Kidd *et al.*, 1989). These RFLPs provide reference points throughout the human genome but only now are they being applied to problems of anthropological interest. However, their potential, and the potential of related techniques, is virtually limitless.

References

Barker, D., Schafer, M. and White, R. (1984). Restriction sites containing CpG show a higher frequency of polymorphism in human DNA. *Cell*, **36**, 131–8.

Bird, A. P. (1986). CpG-rich islands and the function of DNA methylation. *Nature*, **321**, 209–13.

(1989). Two classes of observed frequency for rare-cutter sites in CpG islands. *Nucleic Acids Research*, **17**, 9485.

Bishop, D. T., Williamson, J. A. and Skolnick, M. H. (1983). A model for restriction fragment length distributions. *American Journal of Human Genetics*, **35**, 795–815.

Botstein, D., White, R. L., Skolnick, M. and Davis, R. W. (1980). Construction of a genetic linkage map in man using restriction fragment length polymorphisms. *American Journal of Human Genetics*, **32**, 314–31.

Cooper, D. N., Smith, B. A., Cooke, H. J., Nieman, S. and Schmidt, J. (1985). An estimate of unique DNA sequence heterozygosity in the human genome. *Human Genetics*, **69**, 201–5.

Devor, E. J. (1988). The relative efficiency of restriction enzymes: An update. *American Journal of Human Genetics*, **42**, 179–82.

Drmanac, R. Petrovic, N., Glisin, V. and Crkvenjakov, R. (1988). A calculation of fragment lengths obtainable from human DNA with 78 restriction enzymes: an aid for cloning and mapping. *Nucleic Acids Research*, **14**, 4691–2.

Feder, J., Yen, L., Wijsman, E., Wang, L., Wilkens, L., Schroder, J., Spurr, N., Cann, H., Blumenberg, M. and Cavalli-Sforza, L. L. (1985). A systematic approach for detecting high-frequency restriction fragment polymorphisms using large genomic probes. *American Journal of Human Genetics*, **37**, 635–49.

Helentjaris, T. and Gesteland, R. (1983). Evaluation of random cDNA clones as probes for human restriction fragment length polymorphisms. *Journal of Molecular and Applied Genetics*, **2**, 237–47.

Hofker, M. H., Skraastad, M. I., Bergen, A. A. B., Wapenaar, M. C., Bakker, E., Millington-Ward, A., van Omenn, G. J. B. and Pearson, P. L. (1986). The X-chromosome shows less genetic variation at restriction sites than the autosomes. *American Journal of Human Genetics*, **39**, 438–51.

Human Gene Mapping 10 (1989). Tenth International Conference on Human Gene Mapping. *Cytogenetics and Cell Genetics*, **51**, 1–1148.

Jeffreys, A. J. (1979). DNA sequence variants in the $^{G}\gamma$, $^{A}\gamma$-, δ, and β-globin genes in man. *Cell*, **18**, 1–10.

Jeffreys, A. J. and Flavell, R. A. (1977). A physical map of the DNA regions flanking the rabbit β-globin gene. *Cell*, **12**, 429–39.

Jeffreys, A. J., Royle, N. J., Wilson, V. and Wong, Z. (1988). Spontaneous mutation rates to new length alleles at tandem-repetitive hypervariable loci in human DNA. *Nature*, **332**, 278–81.

Jeffreys, A. J., Wilson, V. and Thein, S.-L. (1985a). Hypervariable 'minisatellite' regions in human DNA. *Nature*, **314**, 67–73.

(1985b). Individual-specific 'fingerprints' of human DNA. *Nature*, **316**, 76–9.

Jeffreys, A. J., Wilson, V., Thein, S.-L., Weathereall, D. J. and Ponder, B. A. J. (1986). DNA 'fingerprints' and segregation analysis of multiple markers in human pedigrees. *American Journal of Human Genetics*, **39**, 11–24.

Kan, Y. W. and Dozy, A. M. (1978). Polymorphism of DNA sequence adjacent to human β-globin structural gene: relationship to sickle mutation. *Proceedings of the National Academy of Sciences USA*, **75**, 5631–5.

Kelly, T. J. and Smith, H. O. (1970). A restriction enzyme from *Haemophilus influenzae*. II. Base sequence of the recognition site. *Journal of Molecular Biology*, **51**, 393–409.

Kidd, K. K., Bowcok, A. M., Schmidtke, J., Track, R. K., Ricciuti, F., Hutchings, G., Bole, A., Pearson, P. and Willard, H. F. (1989). Report of the DNA committee and catalogs of cloned and mapped genes and DNA polymorphisms. *Cytogenetics and Cell Genetics*, **51**, 622–947.

Knowlton, R. G., Brown, V. A., Braman, J. C., Barker, D., Schunam, J. W., Murray, C., Takvorian, T., Ritz, T. and Donis-Keller, H. (1986). Use of highly polymorphic DNA probes for genotype analysis following bone marrow transplantation. *Blood*, **68**, 378–85.

Lai, E., Birren, B. W., Clark, S. M., Simon, M. I. and Hood, L. (1989). Pulsed field gel electrophoresis. *BioTechniques*, **7**, 34–42.

Lange, K. and Boehnke, M. (1982). How many polymorphic marker genes will it take to span the human genome? *American Journal of Human Genetics*, **34**, 842–5.

Linn, S. M. and Roberts, R. J. (1982). *Nucleases*. Cold Spring Harbor Monograph Series, No. 14. New York: Cold Spring Harbor Press.

Meselson, M. and Yuan, R. (1968). DNA restriction enzyme from *E. coli*. *Nature*, **217**, 1110–14.

18   E. J. Devor

Nakamura, Y., Leppert, M., O'Connell, P., Wolff, R., Holm, T., Culver, M., Martin, C., Fujimoto, E., Hoff, M., Kumlin, E. and White, R. (1987). Variable number of tandem repeat (VNTR) markers for human gene mapping. *Science*, **235**, 1616–22.

Nathans, D. and Smith, H. O. (1975). Restriction endonucleases in the analysis and restructuring of DNA molecules. *Annual Review of Biochemistry*, **44**, 273–93.

Pearson, P. L., Bakker, E. and Flavell, R. A. (1982). Considerations in designing an efficient strategy for localizing unique sequence DNA fragments in human chromosomes. *Cytogenetics and Cell Genetics*, **32**, 308.

Razin, A. and Riggs, A. D. (1980). DNA methylation and gene function. *Science*, **210**, 604–10.

Smith, M. O. and Wilcox, K. W. (1970). A restriction enzyme from *Haemophilus influenzae*. I. Purification and general properties. *Journal of Molecular Biology*, **51**, 379–91.

Southern, E. M. (1975). Detection of specific sequences among DNA fragments separated by gel electrophoresis. *Journal of Molecular Biology*, **98**, 503–17.

Swartz, M. N., Trautner, T. A. and Kornberg, A. (1962). Enzymatic synthesis of deoxyribonucleic acid. *Journal of Biological Chemistry*, **237**, 1961–7.

Tuan, D., Biro, P. B., de Riel, J. K., Lazarus, H. and Forget, B. G. (1979). Restriction endonuclease mapping of the human γ-globin gene loci. *Nucleic Acids Research*, **6**, 2519–44.

Wallace, R. B., Petz, L. D. and Yam, P. Y. (1986). Application of synthetic DNA probes to the analysis of DNA sequence variants in man. *Cold Spring Harbor Symposia on Quantitative Biology*, **51**, 257–61.

Weber, J. L. and May, P. E. (1989). Abundant class of human DNA polymorphisms which can be typed using the polymerase chain reaction. *American Journal of Human Genetics*, **44**, 388–96.

Weller, P., Jeffreys, A. J., Wilson, V. and Blanchetot, A. (1984). Organization of the human myoglobin gene. *EMBO Journal*, **3**, 439–46.

Wijsman, E. M. (1984). Optimizing selection of restriction enzymes in the search for DNA variants. *Nucleic Acids Research*, **12**, 9209–26.

Wong, Z., Wilson, V., Jeffreys, A. J. and Thein, S.-L. (1986). Cloning a selected fragment from a human DNA "fingerprint": isolation of an extremely polymorphic minisatellite. *Nucleic Acids Research*, **14**, 4605–16.

Wong, Z., Wilson, V., Patel, I., Povey, S. and Jeffreys, A. J. (1987). Characterization of a panel of highly variable minisatellites cloned from human DNA. *Annals of Human Genetics*, **51**, 269–88.

Wyman, A. R., and White, R. (1980). A highly polymorphic locus in human DNA. *Proceedings of the National Academy of Sciences USA*, **77**, 6754–8.

Youssoufian, M., Kazazian, H. H., Phillips, D. G., Aronis, S., Tsiftis, G., Brown, V. A. and Antonarakis, S. E. (1986). Recurrent mutations in haemophilia A give evidence for CpG mutation hotspots. *Nature*, **324**, 380–2.

2 Nuclear DNA polymorphisms as tools in biological anthropology

JEFFREY ROGERS

Introduction

Genetics has long made an important contribution to biological anthropology. Genetic processes play a part in many areas of bioanthropological research, either explicitly or implicitly. But in most bioanthropological research, they themselves are not the primary concern. Rather, genetic data serve as indicators of biological relationships among individuals or populations. In either case, advances in our ability to analyse the molecular genetics of human and non-human primate populations provide new questions and new opportunities to answer them. In addition, as knowledge of the total genetic constitution or genome of primate species increases, genetics can be integrated into aspects of bioanthropological research that have not made use of genetic data in the past.

Over the last ten years, knowledge of genetic variability in humans and non-human primates has increased dramatically. This progress is primarily attributable to the development of new laboratory techniques that allow direct examination and manipulation of DNA from the cells of almost any organism. Information is now rapidly accumulating about the basic organization of the eukaryotic genome, and especially information about the structure of the human genome (Human Gene Mapping Workshop 10, 1989; HGM10.5, 1990). Along with their desire to understand the essential structural and functional characteristics of the human genetic constitution, molecular biologists have begun to appreciate what biological anthropologists have long known: that one cannot fully understand the fundamental characteristics of the human genome without also developing an understanding of variation within the species and differences among closely related species.

The technical advances in molecular genetics have a number of consequences for biological anthropology. First, it is now possible to examine the molecular genetic basis of variability within and between species in much more detail. Secondly, genes that control a much wider range of phenotypic characters can now be investigated. Rather than

19

being confined to studies of blood groups and proteins, bioanthropologists can conceivably analyse any of the 1000 functional genes that have been mapped in the human genome, or the 4000 total genetic loci (HGM10; HGM10.5). Thirdly, bioanthropologists can begin to investigate the molecular basis of primate morphological adaptation and evolution, and methods now being developed may eventually lead to detailed study of the molecular basis of adaptively significant variation in anatomy or physiology (see pp. 44–6, below).

This chapter reviews some of the recent work concerning within- and between-species variability in the nuclear genome of primates. It has three goals: (1) to review the different types of nuclear DNA polymorphism found within species and the differences observed among species, (2) to illustrate some of the ways such data have been used to investigate primate and human evolution and (3) to suggest areas of future research effort. To avoid confusion, 'polymorphism' will be used only to refer to differences among alleles at a given locus within a species, while 'variation' or 'sequence differences' will be used more generally to indicate differences either within or between species at a given locus.

The literature concerning these issues is very large, and consequently this review selectively focuses on the nature of variability and the application of molecular techniques to specific evolutionary questions. For more comprehensive treatments of molecular evolutionary genetics, see Nei (1987), M. Weiss (1987), K. Weiss (1988, 1989), Hartl and Clark (1989) and Williams (1989). The structure of the primate nuclear genome, basic features of gene structure and function, and details of the molecular mechanisms producing or restricting variation such as mutation and gene conversion are not discussed.

Types of nuclear DNA polymorphism

Variation in nuclear DNA sequences among individuals within a species or between species can take many forms. Large segments of mammalian genomes do not code for protein or functional RNAs, and have no other known function. As a result, much of the nuclear genome is free to undergo mutational change. It appears that natural selection is a tight constraint on only a portion of the total nuclear genome (Nei, 1987; K. Weiss, 1989), though in a genome the size of the human (3 billion base pairs) even small proportions comprise a great deal of sequence.

In presenting a set of categories describing the different types of molecular variability observed, some of the distinctions made between types are somewhat arbitrary. The purpose of this classification is simply to help organize the great amount of data concerning DNA sequence variation into a manageable number of sets that are largely distinct and mostly internally homogeneous. But these categories are also useful because the various types arise at different rates, and differ in other characteristics, such as the probability that a given mutational event will occur more than once in different genetic lineages or that back mutation to the ancestral condition will obscure the evolutionary history of a DNA segment.

As a result of such differences in basic characteristics, some types of variation are better suited for answering particular questions than are other types. For example, it is inappropriate to estimate the length of time since distantly related species diverged by measuring the differences between them at the rapidly mutating 'DNA fingerprinting' loci which are so valuable for comparisons within populations. Since mutations in these hypervariable loci occur quite frequently, any given sequence will become saturated with changes in only a few million years, and consequently the amount of difference between species will not show a linear relationship with time since divergence. The nature of variation at hypervariable loci makes it impossible to determine the number of mutations that have occurred in a given sequence since its divergence from homologous sequences in related species. On the other hand, hypervariable loci are extremely valuable in analyses of local variation within populations. As the nature of molecular variability in primates is more clearly delineated, it will become possible to select the specific genetic loci or types of variation best suited to answering particular questions.

Point mutations

The smallest and simplest changes in DNA consist of point mutations, substitutions of one nucleotide for another. Such changes can be divided into two types, transitions and transversions. Transitions are changes either from one purine to the other (adenine to guanine or guanine to adenine) or from one pyrimidine to the other (cytosine to thymine or the reverse). All other substitutions are transversions. Li *et al.* (1984) used comparisons of pseudogenes, i.e. non-functional duplicates of functional genes, and their functional counterparts to measure the rates of transitions and transversions in nuclear DNA. They found that transitions account for approximately 60% of all substitutions, despite the fact that there are twice as many possible transversions as transitions. This strongly suggests that the rates of all types of point mutations are not equal. Analyses of sequence differences, especially differences between closely related species, should take this into account (Nei, 1987).

Small insertion–deletion mutations

DNA sequences often reveal differences within or between species that consist of insertions or deletions of base pairs. These changes can involve insertion or deletion of a single nucleotide, but larger changes (up to 30 or 40 base pairs) are also common (e.g. Dalgleish *et al.*, 1986; Maeda *et al.*, 1988; Goodman *et al.*, 1989). These changes are not found as frequently as point mutations, especially as polymorphisms within species, but they seem to be more common than the other types of variability listed below. Some DNA sequences, such as long runs of a single nucleotide, are particularly susceptible to this type of mutation.

Hypervariable or VNTR loci

Jeffreys *et al.* (1985) described what they called 'minisatellite' polymorphisms in human DNA. When certain DNA sequences are used as probes in restriction fragment length polymorphism (RFLP) analyses of human nuclear DNA, large numbers of highly variable restriction fragments are revealed (see also Chapters 1 and 4). Several examples of individual, highly polymorphic regions had previously been described (e.g. Wyman and White, 1980; Bell *et al.*, 1982), but the work by Jeffreys *et al.* demonstrated a method for detecting many such hypervariable loci simultaneously.

These minisatellite loci consist of single-copy DNA sequences surrounding multiple, tandemly arranged repeats of a short nucleotide sequence, usually less than 20 base pairs in length. Variation among chromosomes within a population is produced by changes in the number of tandem repeats. The number of repeats can vary from less than a dozen to hundreds, so that there can be a very large number of alleles at a single locus. These alleles differ primarily in the number of repeats, but point mutations can also occur, causing some repeats to differ from the consensus sequence (Jeffreys *et al.*, 1990).

Several of these loci are known to have extraordinarily high mutation rates. One has been shown to produce new alleles at a rate approaching one mutation in every 20 gametes produced by a given individual (Jeffreys *et al.*, 1988). Others have also been demonstrated to undergo frequent mutation. Indeed any particular hypervariable locus that shows a large number of alleles within a single population probably also has a high mutation rate. When a locus has such a high rate of change, the number of alleles at that locus will increase rapidly until a balance is reached between the introduction of new alleles by mutation and the loss of older alleles through genetic drift. In populations of several thousand individuals, the equilibrium value will be a large number of alleles, and the larger the population the larger the equilibrium level of allelic diversity.

These minisatellite loci are the 'DNA fingerprinting' loci that have received much attention (see Weiss and Turner, Chapter 4). Loci with this type of structure have been called VNTR loci by White and his colleagues (Nakamura *et al.*, 1987b), an abbreviation for Variable Number of Tandem Repeats. More than 70 such loci were known to exist in the human genome in 1987 (Nakamura *et al.*, 1987b) and additional examples have been described since then (HGM10.5, 1990). These loci are very useful in several sorts of genetic analyses. Applications include forensics (Gill *et al.*, 1985), genetic linkage and mapping (Nakamura *et al.*, 1987b), population genetics (Balazs *et al.*, 1989), and primate ethology (Weiss and Turner, Chapter 4).

Recently, other hypervariable loci with a different but related structure have been detected. Several research groups have described 'microsatellite' loci (Litt and Luty, 1989; Weber and May, 1989), which also consist of variable numbers of repeat elements, but differ from minisatellites in the length of the repeat unit. Most microsatellites examined so far consist of repeats of the sequence (CA), only two bases long. Each microsatellite locus contains about 10–30 repeat units, and there are tens of thousands of such loci in the human genome (Weber and May, 1989). Nakamura *et al.* (1987a) have described another class of hypervariable loci: 'midisatellites'. These are tandem repeats of a somewhat longer sequence, but the repeat unit occurs many more times, producing a locus hundreds of kilobases in length. Finally, the centromeres of primate chromosomes also contain many repeats of relatively short, alpha-satellite sequences which are known to be polymorphic in repeat number (Willard *et al.*, 1986).

All these classes of hypervariable and alpha-satellite loci can be extremely informative in analyses of genetic differentiation between individuals within a single population. However, mutation rates are so high at these loci that one cannot always assume that two alleles of identical length are in fact copies of the same ancestral allele. Independent, convergent mutation from different ancestral alleles can create new alleles which are so nearly the same length that they may appear identical in routine laboratory comparisons. The rate of mutation, however, is usually low enough so that identity of length among alleles within a single local population is a reliable indication of genealogical relationship. To confirm any inferred relationship, two or more hypervariable loci can be analysed together. But the most rapidly mutating loci cannot be used in all types of population genetic analyses, and are not appropriate for comparisons of the genetic distances among distantly related populations within or between species. In addition to variation in repeat number, changes in single base pairs within repeat units also occur (Jeffreys *et al.*, 1990) and can be used to trace descent. This adds to the precision of genetic studies by allowing more detailed comparisons among alleles.

Small-scale rearrangements

Mammalian species often differ as a result of rearrangements in nuclear DNA larger than those discussed above (i.e. larger than 30–40 bp) and quite different in structure and mechanism from hypervariable loci. Some of these rearrangements are insertions produced by retroposition, the reverse transcription of RNA back into DNA, followed by the insertion of this reverse transcript into the existing genomic DNA of a cell. Such insertions are called retroposons (Weiner *et al.* 1986). Several types of retroposons have been described, including processed pseudogenes, Alu repeats and LINE elements (Vanin, 1985). In addition to retroposons, primate genomes exhibit rearrangements of single-copy DNA that are not reverse transcriptions from RNA, but are produced by insertion, deletion or duplication of single-copy DNA without an RNA intermediate. In most of these cases, the details of the mechanisms causing these rearrangements are not known. Several examples of such rearrangements have been discovered in the α- and β-globin gene clusters (e.g. Zimmer *et al.*, 1980; Marks *et al.*, 1986; Rogers *et al.*, 1991). Several have been found in other regions of the genome (e.g. Verga *et al.*, 1989; Wong *et al.*, 1990).

This category of small-scale rearrangements is somewhat arbitrary because the mechanisms which produce them are not yet understood. It is clear that rearrangements of nuclear DNA involving substantially more than 30 or 40 base pairs, but less than entire functional genes, do occur. It is likely that other mechanisms in addition to reverse transcription produce small-scale rearrangements. This is clear in the contrast between processed pseudogenes which go through an RNA intermediate (Vanin, 1985; Weiner *et al.*, 1986) and the transposition of single-copy DNA described by Wong *et al.* (1990), which does not seem to have involved

RNA. When the mechanisms producing these diverse mutations are understood, it may be better to abandon the concept of small-scale rearrangements and distinguish categories of variability and mutation on the basis of mechanism. But until that is possible, it is reasonable to group these rearrangements together and distinguish them both from those involving only a small number of base pairs, as described above, and from the larger mutations discussed below.

Small-scale rearrangements are particularly significant because they are rare events, and each observed rearrangement is very likely to have occurred only once. Point mutations and the other types of variation described above occur frequently enough that any particular change may have taken place more than once independently. For example, the point mutation leading to sickle-cell anaemia (haemoglobin S) has occurred at least three times in the recent history of *Homo sapiens* (Pagnier *et al.*, 1984; Chebloune *et al.*, 1988). Homoplasy in phylogenetic analyses of DNA sequence data from closely related species (see Miyamoto *et al.*, 1987; Goodman *et al.*, 1989) is sometimes attributed to multiple independent point mutations causing the same change in different evolutionary lineages, or to back mutations. Similarly, several different, independent mutations can produce VNTR alleles with the same number of repeat units. In contrast, small-scale rearrangements seem to be rare enough that most are almost certainly unique events. Furthermore, small-scale rearrangements are unlikely to revert to exactly the ancestral condition present before the mutation.

This inference, that any particular small-scale rearrangement has probably occurred only once, means that these mutations are very useful in evolutionary studies of populations or species. Whenever two chromosomes sampled from a single population, from two distinct populations or even two separate species, are found to carry the same small-scale rearrangement, one can infer that those chromosomes are very likely to be direct lineal descendants from the single chromosome in which the rearrangement occurred. (Strictly speaking, this will apply only to the region immediately surrounding the rearrangement, since recombination among homologous chromosomes will eventually separate and redistribute regions that are physically distant on the same chromosome.) Because of its probable uniqueness, any one small-scale rearrangement is a more useful character in phylogenetic analyses than a single point mutation or other observed molecular difference. This logic applies to the analysis of both within- and between-species variation. Comparison of sequences among rearranged chromosomes can provide information concerning the molecular diversification of a single chromosome within a species, or provide evidence of common ancestry among currently distinct species. The primary limitation of these rearrangements is that only a few of them have been described.

Small-scale rearrangements caused by retroposition

Some small-scale rearrangements are produced by retroposition. A variety of retroposons is known (Vanin, 1985; Weiner *et al.*, 1986), and they give a dramatic indication that the mammalian genome is much more dynamic than was believed only a few years ago. Some retroposons are copies of functional, protein-coding genes. Freytag *et al.* (1984), Tsujibo *et al.* (1985) and Anagnou *et al.* (1984, 1988) describe examples from humans of processed pseudogenes, produced by reverse transcription of a messenger RNA that normally is translated into a protein. Examples have also been described in gorillas and chimpanzees (e.g. Samuelson *et al.*, 1990), and others will undoubtedly be found in many primate species. Some processed pseudogenes are not copies of protein-coding genes but copies of genes that produce functional RNAs. Several families of highly repetitive elements, e.g. LINE1 and Alu repeats, are this type of processed pseudogene. Several examples of specific Alu insertions that are present in one or more primate species and absent in others have been described (Trabuchet *et al.*, 1987; Hixson *et al.*, 1988b; Koop *et al.*, 1989). In addition, researchers have found examples of Alu repeats that are polymorphic within *Homo sapiens*, e.g. the tissue plasminogen-activator locus varies among individuals in the presence or absence of an Alu repeat (Friezner-Degen *et al.*, 1986), and other examples of polymorphic Alu repeats are also known (Matera *et al.*, 1990).

Small-scale rearrangements not caused by retroposition

A number of small-scale rearrangements have been described that do not seem to be retroposons. Examples of such small-scale events are shown in Fig. 2.1. Zimmer *et al.* (1980) observed an insertion of about 1 kilobase of DNA between the β-globin and δ-globin genes of gorillas that is not found in humans, chimpanzees or orangutans, and therefore seems to be the result of a recent mutation in the lineage leading to gorillas. When compared with other hominoids, orangutans have a unique insertion of 200 base pairs 5′ to their α-globin genes and *Hylobates lar* has an extra 900 nucleotides present between the two copies of α-globin. Some of these differences resulting from small-scale rearrangments are

Genomic DNA from Species 1

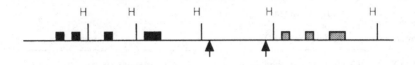

Genomic DNA from Species 2

Fig. 2.1. Illustration of two small-scale rearrangements. This diagram represents a hypothetical region of DNA that contains two loci, one with exons drawn as black rectangles and one with exons drawn as stippled. Species 1 is the ancestral condition and Species 2 the derived. The vertical lines indicate HindIII restriction sites used as landmarks. In Species 2, an insertion has occurred within the first locus and a deletion has occurred between the two loci. The breakpoints of the deletion are indicated by arrows.

shared among species within a monophyletic group. Humans, chimpanzees and gorillas seem to share a 300 bp deletion 3' to the α-globin genes relative to orangutans and gibbons (Zimmer *et al.*, 1980). Four species of gibbons all share an insertion of approximately 700 bp between their pseudo-eta and δ-globin genes (Rogers *et al.*, 1991). Thus this sequence is 700 nucleotides longer in the hylobatids than other hominoids and Old World monkeys.

Some non-retroposon rearrangements are polymorphic within species. Both polymorphic and non-polymorphic rearrangements that have been found in primates are summarized in Table 2.1. The best studied primate species is of course humans, and several examples of human polymorphisms resulting from small-scale rearrangements have been described. Giuffra *et al.* (1990) describe the insertion of part of the fibronectin receptor β-chain gene (FNRB), which is found on human chromosome 10, into chromosome 19. Some but not all copies of human chromosome 19 carry the sequence homologous to FNRB. The locus on chromosome 19 where this insertion occurred has been named the fibronectin receptor beta chain-like locus (FNRBL) and the presence of the FNRBL insertion is polymorphic in all 15 ethnic groups tested to date (Giuffra *et al.*, 1990).

Other types of polymorphic rearrangements have also been reported. The human polymorphism at D1S85 (Verga *et al.*, 1989) is an insertion/deletion polymorphism in which the two alleles differ in size by about 500

Table 2.1. *Small-scale rearrangements in primates*

Species	Insertion (Ins.) or deletion (Del.)	Size (bp)	Locus	Reference
(A) Small-scale rearrangements not known to be polymorphic within species				
Baboon	Ins.	900	α-globin	Marks *et al.*, 1986
Baboon, Rhesus macaque	Ins.	600	APOE	Hixson *et al.*, 1988b
Hylobates lar	Ins.	900	α-globin	Zimmer *et al.*, 1980
Hylobates sp.	Ins.	700	eta-delta globin	Rogers *et al.*, 1991
Orangutan	Ins.	200	α-globin	Zimmer *et al.*, 1980
Human–Chimp– Gorilla	Del.	300	α-globin	Zimmer *et al.*, 1980
Human–Chimp– Gorilla	Both		amylase	Samuelson *et al.*, 1990
Gorilla	Ins.	1000	β/δ globin	Zimmer *et al.*, 1980
Gorilla	Ins.	300	β-globin	Trabuchet *et al.*, 1987
Human	Ins.	1700	ASSP1	Freytag *et al.*, 1984
Human	Ins.	?	LDHA	Tsujibo *et al.*, 1985
(B) Small-scale rearrangements known to be polymorphic within a species				
Rhesus macaque	Ins.	1.8 kb	α-globin	Marks *et al.*, 1986
Human	?	500	D1S85	Verga *et al.*, 1989
Human	Ins.	?	DHFRP1	Anagnou *et al.*, 1988
Human	Ins.	?	FNRBL	Giuffra *et al.*, 1990
Human	Ins.	15 kb	DNF21S2	Wong *et al.*, 1990
Human	Ins.	?	TPA	Friezner-Degen *et al.*, 1986

base pairs. Wong *et al.* (1990) report an unusual type of polymorphism on human chromosome 16. More than 15 kilobases of DNA from chromosome 6 have been inserted into chromosome 16. Like the fibronectin receptor example, some but not all copies of chromosome 16 carry this insertion. All copies of chromosome 6 carry what seems to be the original copy of the sequence. Furthermore, a minisatellite locus is contained within the 15 kb of transposed DNA. As a result, two copies of the same minisatellite, one on chromosome 6 and another on 16, occur in some individuals. The minisatellite locus on chromosome 16 shows a lower level of allelic variation than its paralog on chromosome 6, which suggests that the transposition occurred relatively recently. If the insertion had occurred long ago, mutations would have accumulated in the chromosome 16 copy of the minisatellite and a large number of alleles would be observed. Instead, the chromosome 16 locus exhibits lower levels of variation. Wong *et al.* suggest that there has not been sufficient time to accumulate mutations leading to new alleles, though it is also possible that a recent allelic replacement has occurred near the minisatellite on chromosome 16 and that the number of alleles was reduced by genetic hitch-hiking.

Finally, Anagnou *et al.* (1984, 1988) found that dihydrofolate reductase, which is a functional gene on human chromosome 5, has at least four processed pseudogenes scattered throughout the human genome. One of these pseudogenes, DHFRP1, is on chromosome 18 and is polymorphic. Some copies of chromosome 18 tested have the sequence homologous to the functional DHFR locus and others do not. The frequency of chromosomes carrying DHFRP1 varies in different parts of the world, but the pseudogene is present at some frequency in all five of the populations tested.

Gene duplications and deletions

Gene duplication has been recognized as a major aspect of long-term evolutionary change for many years. In primates, large numbers of genes are members of related groups, or gene families, that exhibit great similarity and presumably were produced by a series of duplications of a single ancestral gene. It may be that a large proportion, or even most, of the functional genes in any given primate species are members of gene families. It is clear from the list of known human genes (Human Gene Mapping Workshop 10, 1989) that many different gene families exist in primates. Maeda and Smithies (1986) and Nei (1987) have provided reviews of this topic.

The existence of a large number of gene families has been documented, but few have been studied in detail, especially in primates other than humans. Without question, the best studied gene families are the α- and β-globin gene clusters (Miyamoto *et al.*, 1987; Savatier *et al.*, 1987; Maeda *et al.*, 1988; Tagle *et al.*, 1988; Goodman *et al.*, 1989; Koop *et al.*, 1989; Fitch *et al.*, 1990; Marks, Chapter 7). Examination of other gene families has begun, but in primates this has not progressed as quickly as the globin studies. The homeobox genes constitute a gene family or set of related gene families that are involved in the regulation of development (Hart *et al.*, 1987; Akam, 1989; Boncinelli *et al.*, 1989). Collagen genes, HLA genes, histone genes, and genes coding for growth hormones, myosin and many other proteins occur as sets of paralogous genes which presumably arose by repeated gene duplication events.

The number of genes in each of these families can vary across species. Zimmer *et al.* (1980) showed that most copies of chimpanzee chromosome 9 (which is homologous to human chromosome 16) contain three copies of the α-globin gene. In other hominoids, the presence of just two functional copies is the rule. Strepsirhines and platyrrhines have only one γ-globin gene, while cercopithecoids and hominoids have two (Koop *et al.*, 1989). Some differences among species are the result of gene deletions rather than duplications. Jeffreys *et al.* (1982) showed that a deletion occurred in the ancestors of modern lemurs that resulted in the fusion of part of the pseudo-eta with part of the δ-globin gene. As a result, modern lemurs have a complement of globin genes different from other primates. Like small-scale rearrangements, such differences are very useful in determining phylogenetic relationships among taxa.

Differences among species in the composition of a given gene family are presumably the result of ancient gene duplications or deletions that occurred in one individual within a species. These events created poly-morphisms within those species that eventually became fixed through genetic drift or natural selection. After a duplication event occurred, and prior to its fixation, different individuals within the species would have differed in gene number just as the descendent species do today. At this point, only a few examples of such polymorphisms within species have been described. The number of functional salivary amylase genes differs among humans (Groot *et al.*, 1989); some human chromosomes 1 carry three functional amylase genes while others have nine functional copies. Evidence for other haplotype arrangements has also been observed. Zimmer *et al.* (1980) found that chimpanzees vary in the number of α-globin genes they carry. Human α-thalassaemia is often caused by deletions of α-globin genes. Within some geographic regions where malaria is endemic, people commonly differ in the number of functional α-globin genes (Hill *et al.*, 1989). The number of genes coding for the green cone pigment in the retina of the eye differs among people (Nathans *et al.*, 1986; Drummond-Borg *et al.*, 1989). Colour blindness in humans is associated with deficiencies in functional retinal pigment genes, which are located on the X chromosome. But even among individuals with normal colour vision, the number of green pigment genes varies. Most 'normal vision' X chromosomes carry two green-pigment genes, while others have as many as five or as few as one.

Other larger molecular changes

There are other types of variation that involve even larger regions of genomic DNA. The analysis of karyotype differences among primate species has provided important evolutionary information (Seuanez *et al.*,

1979; Marks, 1983; Stanyon *et al.*, 1986). These differences in chromosome banding patterns involve inversions or translocations of large blocks of DNA. But the exact nature of these mutations at the nucleotide level, e.g. the molecular characteristics of the breakpoints or fusion points, is unknown.

Finally, other very large insertions or deletions of hundreds of thousands of base pairs of DNA are being studied in detail. For example, Wilkie *et al.* (1991) describe a common polymorphism at the end of the short arm of chromosome 16 in humans. They find that the subtelomeric sequences on this chromosome arm differ in length among individuals, with some chromosomes having regions approximately 260 000 bp longer than others. It is not clear at present whether such large differences are at all common or have any functional consequences, but no obvious differences in function were found in this case.

Summary

The data currently available show that there are several different types of nuclear DNA variability in primates, ranging from single nucleotide substitutions to duplications of entire functional genes and even larger rearrangements. All these types of differences occur both within and between species. These categories of mutation differ in a number of important characteristics in addition to size, such as the rates at which they occur and the probability of parallel or back mutation. As a result, different types of variability are useful for different sorts of analyses. Hypervariable loci are particularly suited for local, micro-evolutionary processes such as tracing the transmission of particular chromosomes within populations. They are also valuable in identifying and distinguishing individuals. Analyses of gene duplications or deletions are more appropriate for macro-evolutionary comparisons among species. Overall, the genomes of primate species are more variable and dynamic than was assumed a few years ago. It is impossible to specify 'the DNA sequence of gene A in species X' or even 'the number and arrangement of functional genes in species X', because a very large proportion of the genome is subject to variability. Just as in studies of morphology, research in the molecular genetics of primates must include quantitative assessment of diverse types of variability.

Distribution of variation within primate species

Modern humans

There is a very large literature describing genetic variation within and between human populations (for reviews see, for example, Mourant *et al.*, 1976; Jorde, 1980; Crawford and Mielke, 1982; Sokal, 1988; K. Weiss, 1988, 1989). Much of this information consists of blood group and protein polymorphisms, and has been used successfully to investigate patterns of variability both in broad geographic regions and in smaller local areas. Analyses of DNA polymorphisms are now adding to the existing body of data, and a small proportion of this new information is summarized here to illustrate the value of molecular analyses.

Analyses of protein and blood group polymorphisms have usually involved the comparison of allele frequencies across a number of local populations or social groups such as villages. The introduction of DNA sequence information provides additional opportunities for analysis. Sequence polymorphisms in nuclear DNA can be analysed individually as single-locus polymorphisms with two or more alleles, very much like traditional protein polymorphisms. Alternatively, when several different polymorphisms are found within a single segment of DNA, they can be analysed together as a unit called a haplotype. If more than one nucleotide is polymorphic in a given DNA segment, then each chromosome in the population will carry a particular combination of 'alleles' at the various polymorphic sites. For example, if in a defined segment there are three different nucleotide positions with two alternative bases observed at each position, then a total of 2^3 or 8 different combinations is possible. Each combination of nucleotides on a single chromosome constitutes a haplotype.

Analysis of haplotypes rather than individual polymorphisms is useful for two reasons. First, the short segment of DNA containing several different polymorphisms can be treated as a single locus with multiple alleles. This will provide more information about inter-individual differences than analysing the different polymorphisms separately. Secondly, simultaneous analysis of several polymorphisms in a single region makes it possible to infer the history of mutational changes there. In such simultaneous analyses, haplotypes at a locus are grouped into clusters by virtue of shared common mutations, uniting haplotypes that share more recent common ancestors (see Fig. 2.2). The result is a 'gene tree' which is a partial reconstruction of the process of diversification in a segment of DNA (see Hartl and Clark, 1989; Long *et al.*, 1990 and references therein). In this way, DNA sequence data allow investigation of the

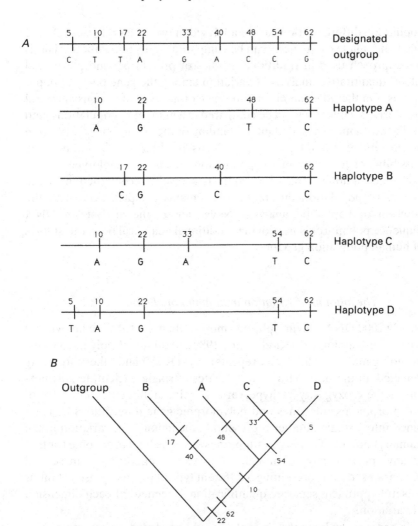

Fig. 2.2. Reconstruction of a 'gene tree' for a hypothetical set of haplotypes at a locus. *A*, The haplotypes with their individual base pair changes from the ancestral sequence. The numbers refer to nucleotide positions within the sequence. *B*, The gene tree based on cladistic analysis of those mutational changes. This analysis indicates that haplotypes C and D share a more recent common ancestor with each other than either does with haplotypes A or B.

history of change in a locus. This leads to a more complete description of evolutionary processes and histories.

With the two approaches to analysing DNA polymorphism (independent study of single polymorphisms and haplotype analysis), human population differentiation can be analysed in two different ways. On the one hand, allele frequencies can be compared across populations. This is the approach used in traditional studies of protein polymorphism and allows quantitative analysis of variation among the gene pools of populations. On the other hand, the presence or absence of various ancestral and derived haplotypes can be compared in a number of populations, and in this way a more detailed understanding of the geographic distribution of clusters of related alleles at a locus is obtained. Of course, the possibility of gene flow among populations means that haplotypes can be exchanged among populations and, as a result, reconstruction of the history of population differentiation remains complex even with the addition of haplotype analysis. Nevertheless, the analysis of DNA sequence polymorphisms provides additional analytical power to studies of human population genetics.

The amount of variation in human populations

Nearly 2000 DNA polymorphisms, most of them RFLPs, are known and have been catalogued (Kidd *et al.*, 1989). Additional polymorphisms, including microsatellites (CA repeats), VNTR loci and others are being detected all the time. How can estimates of single nucleotide polymorphism, heterozygosity at hypervariable minisatellite loci, numbers of polymorphic pseudogenes and polymorphic gene duplications be combined into a single number representing the amount of 'variation in the human genome?' There is no simple answer. The human genome (or that of any species) may be thought of as exhibiting variation in several dimensions, each representing a different type of polymorphism. If this is a useful approach, separate quantification is needed for each dimension of variation.

In order to evaluate the current state of knowledge concerning human molecular variability, it is important to understand how the available information has been obtained. For the most part, human DNA variants are found in the course of studies not specifically designed to measure polymorphism in an unbiased manner. Generally, when a new DNA sequence is cloned, it is tested for polymorphism. This most often involves using the new clone as a probe in RFLP analyses, though other strategies have also been used (Nakamura *et al.*, 1988). The individuals tested for such RFLPs are normally a small group of people of European descent, because so much molecular biological research is done in the United States or Europe. Consequently, large numbers of clones have been tested for polymorphism in small samples essentially from one geographic region. Much less is known about variation in populations from other continents. This is far from ideal for estimating population genetic parameters such as average heterozygosity and levels of population differentiation. The situation is improving, some studies are now under way that will improve our knowledge of average heterozygosity within populations and variability across geographic regions (see, for example, Bowcock *et al.*, 1987; Daiger *et al.*, 1989). The development of broad-based, unbiased estimates of average heterozygosity is unlikely to be a priority for the biomedical research community, and cannot be done using loci which have been pre-selected because they are polymorphic in European populations. Studies of average heterozygosity and regional variability at randomly selected loci should be undertaken.

While no single locus can represent variation in the overall genome, some preliminary estimates based on single loci can provide useful indications. In one of the first such studies, Jeffreys (1979) examined RFLP variation among sixty individuals in the β-globin region. He found that three of the 300 nucleotides falling in restriction sites were polymorphic, and therefore concluded that roughly one out of 100 nucleotides in this chromosomal segment is polymorphic. Ewens *et al.* (1981) showed that simply dividing the number of nucleotides observed to be polymorphic by the total number sampled overestimates the proportion of positions that vary. Using the β-globin data, Ewens *et al.* showed that a better estimate of the proportion of nucleotides polymorphic in that gene cluster is 0.5%. Chakravarti *et al.* (1984) surveyed RFLP polymorphism in the human growth hormone gene cluster among 100 chromosomes. They found less polymorphism than in the β-globin analyses above; approximately one out of every 500 nucleotides was variable. Analyses using RFLP, hypervariable loci and denaturing gradient gel electrophoresis of single-copy sequences (Lerman *et al.*, 1984; Myers *et al.*, 1985; Abrams *et al.*, 1990) will ultimately produce a general picture of the amount of nuclear DNA polymorphism in human populations. But reliable estimates will come only from studies designed to measure variation in a broad and unbiased manner.

Studies of genetic differentiation among human populations

One of the goals of bioanthropological research is the reconstruction of the history of dispersal and differentiation of modern human populations (see, for example, Suarez *et al.*, 1985; K. Weiss, 1988; Long *et al.*, 1990;

Chen *et al.*, 1990). Such analysis requires accurate characterization of the genetic composition of the groups under study, and molecular genetics now provides the opportunity to examine genetic differentiation in much greater detail than has been possible in the past. The following discussion includes a few selected examples of both (a) comparisons of allele frequencies across population samples and (b) examination of the differentiation of haplotypes at a given locus. In a few cases, these approaches are combined in a single study.

Bowcock *et al.* (1987) examined 47 different RFLP systems in five samples from geographically separated human groups: Europeans, Chinese, Biaka Pygmies, Mbuti Pygmies and Nasioi Melanesians. They found that the great majority of restriction site polymorphisms originally described in Europeans also exist in the other four groups. Even the Nasioi sample, which consisted of about 20 unrelated individuals sampled from a single locality on Bougainville Island, showed substantial intra-population variability. Each of the five groups does have unique 'private' polymorphisms, i.e. alleles found only in that group. But most alleles were found in all five samples. These studies are being extended to other ethnic groups from Asia, Africa and the Americas, in order to describe more completely the distribution of alleles at these loci (Kidd and Kidd, 1990).

Wainscoat *et al.* (1986) examined sequence variation in the β-globin gene cluster, testing over 600 chromosomes sampled from ten geographically distributed populations for five different RFLP marker systems. The majority of chromosomes examined carry one of five haplotypes, and the amount of difference between these haplotypes suggests that some have been undergoing separate evolution for a long period of time. The most common haplotype of the five found among the 61 chromosomes studied from African samples was not present in a total of 540 haplotypes from non-African populations. Additional haplotypes, which are probably recent minor variants of the older haplotypes, also occurred in the non-African sample.

The key elements of these data are (1) that there are differences in haplotype frequencies between African and all non-African populations tested and (2) that the variation outside Africa seems to be a subset of the variation within Africa. Wainscoat *et al.* (1986) cautiously suggest that this is consistent with the notion that modern humans evolved in Africa, and that later some population(s) migrated from that continent to diversify throughout Eurasia. The differences between African and non-African populations are attributed to genetic drift and founder effects which occurred during and immediately subsequent to the migration from Africa to Eurasia. This study illustrates the increased power of DNA sequence polymorphisms over protein variation. Wainscoat *et al.* (1986) were able to detect several alleles (haplotypes) at a single locus, to compare the frequencies of those alleles in various populations, and in some cases to distinguish ancestral from derived haplotypes.

The conclusions reached by Wainscoat *et al.* (1986) have been both endorsed and criticized. Jones and Rouhani (1986) concurred with the original interpretation and even extended the analysis, making several assumptions and then calculating the amount of genetic drift that took place at the time of dispersal from Africa. This results in an estimate of the effective population size of the population that migrated from Africa to Eurasia. Giles and Ambrose (1986) correctly pointed out that this interpretation is not the only one compatible with the data, and that a Eurasian origin for modern humans is not ruled out. Data subsequently published by Cann *et al.* (1987) and Vigilant *et al.* (1989) regarding mitochondrial DNA add further support to the original analyses of Wainscoat *et al.* (1986).

Nuclear DNA polymorphisms have also been applied to the study of biological relationships among populations in the Pacific region (see Flint *et al.*, Chapter 5). This work focused on the α- and β-globin gene clusters and the many mutations observed in them, including those which produce α- and β-thalassaemia (Hill *et al.*, 1989). Using both polymorphism in single nucleotide sites and the more informative thalassaemia-causing deletion mutations, Hill and colleagues reconstructed the probable relationships among populations in Melanesia, Polynesia and Micronesia and inferred likely routes of migration across Melanesia into Polynesia and Micronesia (Hill and Serjeantson, 1989).

Ramsay and Jenkins (1987, 1988) investigated DNA haplotype polymorphisms in several African groups. Their data on the globin gene clusters in San and Bantu-speaking populations in southern Africa support the notion that the β-globin haplotype polymorphism is an old one (Wainscoat *et al.*, 1986) and indicate that the San and Bantu-speakers have significantly different frequencies of α- and β-globin haplotypes. There is a rich history of population differentiation (and admixture) in Africa as well as Eurasia (see also Nurse *et al.*, 1985; Excoffier *et al.*, 1987).

A different segment of the genome, the pseudoautosomal region, is a sequence of approximately one million base pairs at the distal tip of the short arm of human X and Y chromosomes (Ellis and Goodfellow, 1989). Ellis *et al.* (1990) examined polymorphisms in the pseudoautosomal boundary region, where the homology between X and Y chromosomes ends and the sequences become sex-chromosome specific. There are several polymorphisms in the first 300 bp of pseudoautosomal sequence adjacent to the boundary. Among 57 Y and 60 X chromosomes tested, originating from ten different human populations, six nucleotide sites were polymorphic. Three sites were polymorphic only on X chromosomes, one site was variable on both X and Y chromosomes and two positions were fixed for different nucleotides on the X and Y. Several haplotypes differing by more than one mutation were found on X chromosomes segregating in European, Asian and African populations. All Y chromosomes from non-African samples carried the same 300 bp sequence, while the only variation in this segment among Y chromosomes was found in African populations.

Two conclusions follow from these data. First, the variation found in non-African populations is a subset of the variability observed in African samples. Two different sequences were found on Y chromosomes sampled from Africa, and only one of them was present in Eurasia. Eight haplotypes were found among 19 African X chromosomes and only five of them were observed in a larger Eurasian sample. This is the same basic pattern of diversity and geographic differentiation found for RFLP haplotypes in the β-globin cluster (Wainscoat et al., 1986), and is further support for the idea that modern Eurasian groups are derived from ancestors who lived in Africa.

The second conclusion results from comparing levels of diversity among X chromosome sequences to diversity among Y-sequences. Only one nucleotide demonstrated any variation in the 300 nucleotide segment among 57 Y chromosome segments. In an equivalent number of nucleotides from 60 X chromosomes, five sites were variable. Since there is no reason to believe that the mutation rate is higher on X than Y chromosomes, the best explanation for the larger variability among X's is that the X chromosomes have been diversifying for a longer period (Ellis et al., 1990; Maynard Smith, 1990). In other words, the last common ancestor of all extant Y-chromosome pseudoautosomal boundaries existed at a time later than the last common ancestor of all observed X-chromosome boundaries. This illustrates (a) that the evolutionary history of any particular DNA segment can be inferred from the pattern of variability it exhibits and (b) that the inferred evolutionary history applies to that segment only. Other segments of DNA may have different histories of differentiation, loss of variation and subsequent re-diversification. We can construct independent 'gene trees' for different segments of nuclear DNA, but not all segments will present the same tree because they will not all have the same history of differentiation. This has implications for attempts to reconstruct both the history of dispersal in *Homo sapiens* and the phylogeny of closely related species such as humans and African apes (Pamilo and Nei, 1988).

Inferences about natural selection

One of the processes that exerts significant influence over the nature and amount of molecular genetic variation is natural selection. In the past, practical limitations have made detection of selection in human populations extremely difficult. It is still difficult. But because it is now possible to examine a larger number of genetic loci in much more detail, more complete descriptions of the genetic differences among individuals can be developed. This will eventually lead to a better understanding of all the processes that affect variation, including selection.

With the exception of haemoglobinopathies and malaria, there are still no definitive examples of natural selection in human populations, though other strong candidates have been proposed. Weiss *et al.* (1984) gave one example related to nutritional stress. Geographic correlations between the frequencies of specific α-globin mutations and the incidence of malaria indicate that there is a clear relationship between the disease and thalassaemia mutations in Melanesia and surrounding areas (Hill *et al.*, 1989; Flint *et al.*, Chapter 5). Recent molecular studies of β-globin variation in Africa have provided new insight into the history of malaria-induced selection there. Kan and Dozy (1980) first showed an association between specific alleles for restriction fragment length polymorphisms near the β-globin gene and mutations causing sickle-cell anemia. They found that in different areas of Africa, of the Mediterranean region and of Asia, sickle-cell mutations (HbS) were found on chromosomes with different restriction site patterns, i.e. different haplotypes. Their explanation was that the HbS mutation had occurred more than once, each time on a chromosome with a different restriction site haplotype, and that these different mutation-bearing chromosomes had independently increased in frequency because of selection for the HbS mutations. Subsequent study (Pagnier *et al.*, 1984; Chebloune *et al.*, 1988) has confirmed this hypothesis. Pagnier and colleagues determined the β-globin multi-site haplotypes for 124 HbS chromosomes from four different African populations. The predominant haplotype found on HbS chromosomes from Benin differed from that found in the Central African Republic, and chromosomes from Senegal showed primarily a third type. The three haplotypes differed by several restriction sites and thus the best explanation is that the same HbS mutation has occurred at least three times. Other processes may also be involved in the production of these haplotypes, such as gene conversion. But however it arose, the new HbS chromosome in each region eventually increased in frequency independent of allele frequencies in the other areas.

While malaria and haemoglobinopathies provide the strongest evidence for selection in human populations, other possible examples are available. Among European populations, phenylketonuria (PKU) is one of the most common genetic diseases. It is caused by recessive mutations that result in deficiency of the enzyme phenylalanine hydroxylase, and the proportion of carriers for these mutations reaches 1 in 50 in some groups (Scriver and Clow, 1980). Molecular analysis of PKU mutations (DiLella et al., 1987; Daiger et al., 1989) established that PKU mutations have occurred on numerous different chromosomal backgrounds. In some cases, haplotypes that are rare among normal chromosomes are quite common among PKU chromosomes. Kidd (1987) suggested that high mutation rates cannot account for the incidence of PKU in these populations, and that genetic drift is unlikely to increase the frequency of more than one independent mutation at the same locus – especially when those mutations cause severe pathology in homozygotes. The remaining possibility is that some form of heterozygote advantage has increased the frequency of PKU-associated alleles despite the problems that develop in people homozygous for those alleles. At this point, there is no indication of the selective agent involved (Kidd, 1987), but as data accumulate describing further new PKU alleles that have also increased in frequency (Daiger et al., 1989), the hypothesis acquires additional support.

Other species

Most of the work on molecular genetic variation within non-human primate species has concerned mitochondrial DNA (see Melnick et al., Chapter 6). One study of nuclear DNA polymorphisms in free-ranging yellow baboons (Papio hamadryas cynocephalus) from Mikumi National Park, Tanzania (Rogers, 1989; J. Rogers and K. K. Kidd, unpublished data) found substantial amounts of DNA polymorphism within this single local population. Four loci were tested for RFLPs using human DNA clones as probes. All four exhibited variation. Based on this preliminary sample, it was estimated that between 1 and 2% of randomly sampled nucleotides are polymorphic within this one baboon population. Furthermore, the frequencies of particular RFLP alleles were significantly different in the various social groups sampled. This indicates that nuclear

DNA polymorphisms are not only common in baboons, but that they can be used to investigate patterns of population genetic structure and local demographic processes.

Information is also accumulating concerning DNA polymorphisms in captive primates. There are RFLPs in the low density lipoprotein receptor gene of captive baboons (Hixson *et al.*, 1989) and in the gene for apolipoprotein A-I (Hixson *et al.*, 1988a). Lu *et al.* (1990) found several DNA polymorphisms in the glycophorin genes of chimpanzees (*Pan troglodytes*), while Ely and Ferrell (1990) used human probes to detect hypervariable loci in the same species. The RFLP approach has also been used to study single-copy nuclear DNA polymorphism in rhesus macaques, gorillas and orangutans (Rogers *et al.*, 1991; J. Rogers and K. K. Kidd, unpublished data). Taken together, the results from the RFLPs and hypervariable loci indicate that non-human primate species contain substantial amounts of intraspecific variation, and that human DNA clones can be used as probes to detect this variation.

Other techniques are available which detect single base-pair changes that do not occur in restriction enzyme recognition sites. Denaturing gradient gel electrophoresis has been used to screen a homeobox-containing locus (HOX2B) for within- and between-species variation (Ruano *et al.*, 1990; Rogers *et al.*, 1991). Both gorillas and chimpanzees exhibit within-species variability for this DNA segment. Denaturing gradient gel electrophoresis holds much promise for studies of nuclear DNA polymorphism in humans and non-human primates because it can detect single nucleotide changes which do not occur within restriction sites, and thus is more sensitive than RFLP analysis (Lerman *et al.*, 1984; Myers *et al.*, 1985; Abrams *et al.*, 1990).

Variation among primate species

Most molecular comparisons among primate species have been made to determine phylogenetic relationships and reconstruct evolutionary histories of specific taxonomic groups. A great deal of effort has been focused on interspecies comparisons of the α- and β-globin gene families, reviewed by Koop *et al.* (1989), Goodman *et al.* (1989), Maeda *et al.* (1988), Savatier *et al.* (1987), Nei (1987) and Marks (Chapter 7). Among haplorhine and strepsirhine primates, all the types of DNA sequence variation described above have been observed. Indeed, a very large proportion of our knowledge of molecular variability in the nuclear genomes of primates comes from the study of these two gene clusters. Such comparisons based on these gene clusters have provided important information concerning the phylogeny of primates, but have also helped in documenting molecular aspects of changes in physiological and developmental processes through primate evolution (e.g. Tagle *et al.*, 1988).

As human molecular genetics progresses, more studies are including interspecific comparisons as an aid in the interpretation of human molecular data. Recent studies have examined interspecies differences in ribosomal genes (Gonzalez *et al.*, 1990), albumin genes (Murray *et al.*, 1984), involucrin (Djian and Green, 1989), apolipoprotein genes (Hixson *et al.*, 1988b) and others. It is not possible to review the results of phylogenetic or developmental analyses here. Rather, a specific aspect of molecular phylogenetic analysis – the implications of intraspecies polymorphism for cross-species comparisons – will be discussed.

The phylogenetic relationships among gorillas, chimpanzees and humans are the subject of much debate. Understanding of hominoid phylogeny has benefited greatly from molecular analyses (e.g. Sarich and Wilson, 1967; Goodman *et al.*, 1989; Caccone and Powell, 1989). Though some authors have concluded otherwise, it is not yet possible to resolve definitively the relationships among the genera *Pan*, *Gorilla* and *Homo*. The current consensus among morphologists is that cranial and postcranial anatomical characters link chimpanzees and gorillas together, with hominids as their sister taxon (Andrews, 1987; Andrews and Martin, 1987). However, some recent molecular studies suggest that humans and chimpanzees are the most closely related of the three genera (Goodman *et al.*, 1989; Caccone and Powell, 1989, and others). Most of the recent debate concerning hominoid phylogeny is an attempt to determine which of these two phylogenies is the correct one.

The genetic data are not consistent on this question. Studies of karyotypes have not reached a consensus (compare Yunis and Prakash, 1982 with Marks, 1983). Goodman and his colleagues concluded that the sequences of globin genes place chimpanzees closer to humans than to gorillas (Goodman *et al.*, 1989). But the published globin data indicate that (1) there are individual nucleotide changes that suggest the other two possible phylogenies are correct, i.e. there is homoplasy (see Goodman *et al.*, 1989) and (2) while sequence differences in eta-globin suggest a chimpanzee–human clade, the β-globin locus actually suggests that the human lineage is the outgroup (Koop *et al.*, 1989). Djian and Green (1989) have sequenced the involucrin gene in *Homo*, *Pan*, *Gorilla* and *Pongo*, and found 10 derived characters shared by *Pan* and *Gorilla* while none link *Pan* and *Homo*. The mitochondrial genome, which evolves more rapidly than the nuclear, has not provided a clear picture of the phylogeny (see review by M. Weiss, 1987).

The DNA–DNA hybridization approach (e.g. Caccone and Powell, 1989) was thought to have answered the question, but it has not received universal acceptance. Work in various laboratories showed that these hybridization techniques provide reliable information about molecular differences between DNA sequences (Caccone *et al.*, 1988). The hybridization studies of hominoid taxa have been challenged on a number of grounds (Andrews, 1987; Marks *et al.*, 1988; Sarich *et al.*, 1989; Schmid and Marks, 1990). Even the most persistent critics of DNA–DNA hybridization as an approach to phylogenetic analysis agree that one can obtain information about divergence between DNA molecules under some circumstances (Schmid and Marks, 1990). At issue is the reliability of DNA–DNA hybridization experiments for determining the phylogenetic relationships among very closely related species. At this point it is best to conclude that the method is valuable for measuring differences among individual DNA molecules, and that it is useful for phylogenetic analyses among taxa that have diverged substantially, but that its ability to discern the branching sequence among very closely related mammalian taxa has not yet been definitively established.

Rather than arguing about which method or locus is providing the one true phylogeny, the consequences of polymorphism within primate species for between-species comparisons must be considered. Clearly, there are substantial amounts of several different types of variation segregating in primate populations. If the last common ancestor of *Homo*, *Pan* and *Gorilla* was polymorphic at many different loci, then analysis of different loci in the descendent taxa can result in different gene trees and thus different phylogenies (Nei, 1986, 1987; Smouse and Li, 1987; Maeda *et al.*, 1988; Pamilo and Nei, 1988). Which phylogeny is suggested by any given locus depends on how the various polymorphic alleles segregating in the ancestor were lost or retained in the descendent lineages. This means that if an ancestral species is polymorphic, then its daughter species can be mosaics of various combinations of the genetic variation which was present in the ancestor. It may be true that at some loci chimpanzees are more similar to humans than to gorillas *and* true that the chimpanzees are more similar to gorillas than to humans at other loci. In fact, depending upon the level of polymorphism in the ancestor and the time between the two phylogenetic branching events, this result may be highly likely (Nei, 1986; Pamilo and Nei, 1988).

If viewed in this way, the conflict among data concerning hominoid phylogeny is not a problem to be resolved by accepting some and rejecting others. Rather, the inconsistencies may be telling us something about the biological processes involved in the population differentiation and radiation of the last common ancestor into three independent lineages. There is one true historical relationship among the lineages of organisms that now constitute humans, chimpanzees and gorillas, even if individual genes are not always reliable indicators of it. A complete analysis must include assessment of the differences between species in the light of emerging patterns of within-species variation.

Future directions for research

Testing models of population structure

Several different approaches have been used to describe and explain the distribution of genetic variation across human populations (see Jorde, 1980 and Hartl and Clark, 1989 for discussions of the theory and Crawford and Mielke, 1982 and K. Weiss, 1989 for representative empirical analyses). The theoretical models used fall into several classes. Some assume migration occurs between subdivisions of large populations, i.e. island models and stepping stone models. Other approaches (e.g. Nei and Roychoudhury, 1982; Cavalli-Sforza *et al.*, 1988) have assumed that essentially no migration occurs between most isolates once they diverge. These various models represent fundamentally different conceptual approaches to the analysis of human population genetic structure, and different models will be appropriate in different circumstances. One challenge for the future is to determine which models are most appropriate for specific empirical cases.

The new forms of nuclear DNA data (complex DNA sequence haplotypes, hypervariable loci, rare and unique rearrangements of DNA, etc.) will provide much more detailed genetic information for testing the validity of these models in specific circumstances. Furthermore, new theoretical models and analytical tools are now being developed for these purposes (Slatkin and Maddisson, 1989, 1990). The incorporation of both mitochondrial and nuclear DNA studies into analyses of local and regional population structure will help in assessing the utility of the alternative models. Bioanthropologists have already begun to contribute other new analytical approaches based on the molecular haplotype data (e.g. Long *et al.*, 1990).

Investigating the genetic basis of primate adaptation

The study of adaptive evolutionary change has always been a major part of bioanthropology, but human and non-human primate genetics has with few exceptions been the study of neutral genetic markers. The rapid progress in molecular biology over the last few years has generated new approaches for the study of mammalian embryology and development. Biologists are now making tremendous progress in isolating and analysing genes that are involved in the control of developmental processes in mammals (see, for example, Boncinelli *et al.*, 1989; Kessel *et al.*, 1990). Some of the genes currently under study influence traits of particular interest to bioanthropology. HOX5 is a cluster of genes which are expressed in the limb bud during critical periods of embryonic growth (Dolle *et al.*, 1989a). Retinoic acid and the proteins that act as its receptors seem to be basic elements in several ontogenetic processes (Dolle *et al.*, 1989b; Kurst *et al.*, 1989). HOX2 and HOX3 are clusters of genes expressed, among other places, in the developing central nervous system (Holland and Hogan, 1988; Akam, 1989; Gaunt *et al.*, 1989). It is premature to speculate about the genes which influence the differences between the forelimbs of gibbons, baboons and tarsiers, or other anatomical systems in other species. But mammalian developmental genetics is a field in which dramatic advances will be made, and those advances will have consequences for the study of primate morphological evolution. Documenting molecular variability within and between species at these genetic loci, and then correlating observed differences with observed morphological differences, will be an important part of this line of research. Bioanthropologists can take a leading role in such investigations.

One way to begin is to examine the genetic basis of quantitative phenotypic variation in traits of interest to anthropology. For example, Cheverud *et al.* (1990) examined the heritability of several features of external brain morphology in rhesus monkeys. This approach can demonstrate which phenotypic characters are most amenable to genetic analysis. Then, analytical procedures can be used to locate in the genome previously unknown genes that influence phenotypic variation. Weiss (1989) presented an introduction to genetic linkage analysis as it applies to human gene mapping. This approach has been successful in detecting and isolating genes involved in various human diseases, and can be applied to a wider range of research questions. Specific methods also exist for detecting and mapping genes that control quantitative phenotypic variation (Boerwinkle *et al.*, 1986; Lander and Botstein, 1989; Blangero *et al.*, 1991). Such analyses are already being used in studies of biochemical and physiological variation among non-human primates (MacCluer *et al.*, 1988; Hixson *et al.*, 1989; Blangero *et al.*, 1989). Bioanthropology would benefit greatly from the eventual integration of molecular, genetic, morphological and developmental analyses of primate evolution.

46   J. Rogers

Conclusions

A number of different types of molecular variability exist within and between primate species. Recent research has uncovered a wide variety of mutational changes in primate nuclear DNA, and although little work has been done in species other than *Homo sapiens*, it is possible to draw some general conclusions. The types of molecular differences detected within and between species can be provisionally subdivided into a small number of categories, though the distinctions between several of these categories are somewhat arbitrary. Advances in knowledge of the molecular mechanisms that cause these mutations will probably necessitate changes in these categories. In any case, the different types of variation described above are useful for different types of analyses. Some genetic loci mutate rapidly and are well suited for analysis of the distribution of variation in local or regional populations within a single species. Other types of differences accumulate slowly and are better suited for long-term macroevolutionary comparisons between species or higher taxonomic groups.

More is known about variability in humans than in other species and this will probably always be the case, given the role of molecular genetics in research related to human disease. Nevertheless, researchers are now examining variability within other species, and anthropology will benefit widely from increased knowledge of this molecular variation. Human population genetics is an area of continuing rapid progress. Nuclear DNA data are also being used in studies of phylogenetic relationships among species, and this work will expand as researchers compare more species and additional loci in the species already investigated. In the future, it should be possible to expand the use of nuclear DNA data in studies of population genetics, population structure and analyses of the molecular basis of primate morphological evolution.

Acknowledgements
I wish to thank Dr Eric Devor for the opportunity to contribute to this volume. This chapter draws heavily on the training and experience I gained while working with Dr Kenneth K. Kidd. I am grateful to him for that opportunity, for the guidance and resources he has provided to me, and for his comments on an early draft of this paper. I also thank Drs Alison Richard and Luis Giuffra for their helpful comments on this paper. While writing this chapter, the author was supported by NSF grant BNS8813234 to Kenneth K. Kidd.

References
Abrams, E. S., Murdaugh, S. E. and Lerman, L. S. (1990). Comprehensive detection of single base changes in human genomic DNA using denaturing gradient gel electrophoresis and a GC clamp. *Genomics*, **7**, 463–75.

Akam, M. (1989). Hox and Hom: Homologous gene clusters in insects and vertebrates. *Cell*, **57**, 347–9.

Alberts, B., Bray, D., Lewis, J., Raff, M., Roberts K., and Watson J. D. (1983). *Molecular Biology of the Cell*. New York: Garland Publishing, Inc.

Anagnou, N. P., O'Brien, S. J., Shimada, T., Nash, W. G., Chen, M.-J. and Nienhuis, A. W. (1984). Chromosomal organization of the human dihydro-folate reductase genes: Dispersion, selective amplification and a novel form of polymorphism. *Proceedings of the National Academy of Sciences USA*, **81**, 5170–4.

Anagnou, N. P., Antonarakis, S. E., O'Brien, S. J., Modi, W. S. and Nienhuis A. W. (1988). Chromosomal localization and racial distribution of the polymor-phic human dihydrofolate reductase pseudogene (DHFRP1). *American Journal of Human Genetics*, **42**, 345–52.

Andrews, P. (1987). Aspects of hominoid phylogeny. In *Molecules and Mor-phology in Evolution: Conflict or Compromise*, ed. C. Patterson. Cam-bridge: Cambridge University Press.

Andrews, P. and Martin, L. (1987). Cladistic relationships of extant and fossil hominoids. *Journal of Human Evolution*, **16**, 101–18.

Balazs, I., Baird, M., Clyne, M. and Meade, E. (1989). Human population genetic studies of five hypervariable DNA loci. *American Journal of Human Genetics*, **44**, 182–90.

Bell, G. I., Selby, M. J. and Rutter, W. J. (1982). The highly polymorphic region near the human insulin gene is composed of simple tandemly repeating sequences. *Nature*, **295**, 31–5.

Blangero, J., Kammerer, C., Konigsberg, L., Hixson, J. and MacCluer, J. (1989). Statistical detection of genotype–environment interaction: A multi-variate measured genotype approach. *American Journal of Human Genetics*, **45**, A234.

Blangero, J., Williams-Blangero, S. and Hixson, J. E. (1991). Assessing the effects of candidate genes on quantitative traits in primate populations. *American Journal of Primatology* (in press).

Boerwinkle, E., Chakraborty, R. and Sing, C. F. (1986). The use of measured genotype information in the analysis of quantitative phenotypes in man. I. Models and analytical methods. *Annals of Human Genetics*, **50**, 181–94.

Boncinelli, E., Acampora, D., Panese, M., D'Esposito, M., Somma, R., Gaudino, G., Stornaiuolo, A., Cafiero, M., Faiella, A. and Simeone, A. (1989). Organization of human class I homeobox genes. *Genome*, **31**, 745–56.

Bowcock, A. M., Bucci, C., Hebert, J. M., Kidd, J. R., Kidd, K. K., Fried-laender, J. and Cavalli-Sforza, L. L. (1987). Study of 47 DNA markers in five populations from four continents. *Gene Geography*, **1**, 47–64.

Caccone, A. and Powell, J. R. (1989). DNA divergence among hominoids. *Evolution* **43**, 925–42.

Caccone, A., DeSalle, R. and Powell, J. R. (1988). Calibration of the change in thermal stability of DNA duplexes and degree of base pair mismatch. *Journal of Molecular Evolution*, **27**, 212–16.

Cann, R. L., Stoneking, M. and Wilson, A. C. (1987). Mitochondrial DNA and human evolution. *Nature*, **325**, 31–6.

48   J. Rogers

Cavalli-Sforza, L. L., Piazza, A., Mendozzi, P. and Mountain, J. (1988). Reconstruction of human evolution: Bringing together genetic, archeological and linguistic data. *Proceedings of the National Academy of Sciences USA*, **85**, 6002–6.

Chakravarti, A., Phillips, J. A. III, Mellits, K. H., Buetow, K. H. and Seeburg, P. H. (1984). Patterns of polymorphism and linkage disequilibrium suggest independent origins of the human growth hormone gene cluster. *Proceedings of the National Academy of Sciences USA*, **81**, 6085–9

Chebloune, Y., Pagnier, Y. J., Trabuchet, G., Faure, C., Verdier, G., Labie, D. and Nigon, V. (1988). Structural analysis of the 5′ flanking region of the beta-globin gene in African sickle cell anemia patients: Further evidence for three origins of the sickle cell mutation in Africa. *Proceedings of the National Academy of Sciences USA*, **85**, 4431–5.

Chen, L. Z., Easteal, S., Board, P. G. and Kirk, R. L. (1990). Evolution of beta-globin haplotypes in human populations. *Molecular Biology and Evolution*, **7**, 423–37.

Cheverud, J. M., Falk, D., Vannier, M., Konigsberg, L., Helmcamp, R. C. and Hildebolt, C. (1990). Heritability of brain size and surface features in rhesus macaques (*Macaca mulatta*). *Journal of Heredity*, **81**, 51–7.

Crawford, M. H. and Mielke, J. H. (ed.) (1982). *Current Developments in Anthropological Genetics. Volume 2: Ecology and Population Structure*. New York: Plenum Press.

Daiger, S. P., Reed, L., Huang, S.-S., Zeng, Y.-T., Wang, T., Lo, W. H. Y., Okano, Y., Hase, Y., Fukuda, Y., Oura, T., Tada, K. and Woo, S. L. C. (1989). Polymorphic DNA haplotypes at the phenylalanine hydroxylase (PAH) locus in Asian families with phenylketonuria. *American Journal of Human Genetics*, **45**, 319–24.

Dalgleish, R., Williams, G. and Hawkins, J. R. (1986). Length polymorphism in the pro-alpha2(I) collagen gene: an alternative explanation in a case of Marfan syndrome. *Human Genetics*, **73**, 91–2.

DiLella, A. G., Marvit, J., Brayton, K. and Woo, S. L. C. (1987). An amino acid substitution in phenylketonuria is in linkage disequilibrium with DNA haplotype 2. *Nature*, **327**, 333–6.

Djian, P. and Green, H. (1989). Vectorial expansion of the involucrin gene and the relatedness of the hominoids. *Proceedings of the National Academy of Sciences USA*, **86**, 8447–51.

Dolle, P., Izpisua-Belmonte, J.-C., Falkenstein, H., Renucci, A. and Duboule, D. (1989a). Coordinate expression of the murine *Hox-5* complex homeobox-containing genes during limb pattern formation. *Nature*, **342**, 767–72.

Dolle, P., Ruberte, E., Kastner, P., Petkovich, M., Stoner, C. M., Gudas, L. J. and Chambon, P. (1989b). Differential expression of genes encoding alpha-, beta- and gamma-retinoic acid receptors and CRABP in the developing limb of the mouse. *Nature*, **342**, 702–5.

Drummond-Borg, M., Deeb, S. S. and Motulsky, A. G. (1989). Molecular patterns of X chromosome-linked color vision genes among 134 men of European ancestry. *Proceedings of the National Academy of Sciences USA*, **86**, 983–7.

Ellis, N. and Goodfellow, P. N. (1989). The mammalian pseudoautosomal region. *Trends in Genetics*, **5**, 406–10.

Ellis, N., Taylor, A., Bengtsson, B. O., Kidd, J., Rogers, J. and Goodfellow, P. (1990). Population structure of the human pseudoautosomal boundary. *Nature*, **344**, 663–5.

Ely, J. and Ferrell, R. E. (1990). DNA fingerprints and paternity ascertainment in chimpanzees (*Pan troglodytes*). *Zoo Biology*, **9**, 91–8.

Ewens, W. J., Spielman, R. S. and Harris, H. (1981). Estimation of genetic variation at the DNA level from restriction endonuclease data. *Proceedings of the National Academy of Sciences USA*, **78**, 3748–50.

Excoffier, L., Pellegrini, B., Sanchez-Mazas, A., Simon, C. and Langaney, A. (1987). Genetics and history of sub-Saharan Africa. *Yearbook of Physical Anthropology*, **30**, 151–94.

Fitch, D. H. A., Mainone, C., Goodman, M. and Slightom, J. L. (1990). Molecular history of gene conversions in the primate fetal gamma-globin genes. *Journal of Biological Chemistry*, **265**, 781–93.

Freytag, S. O., Bock, H., Beaudet, A. L. and O'Brien, W. E. (1984). Molecular structures of human arginosuccinate synthetase pseudogenes. *Journal of Biological Chemistry*, **259**, 3160–6.

Friezner-Degen, S. J., Rajput, B. and Reich, E. (1986). The tissue plasminogen-activator gene. *Journal of Biological Chemistry*, **261**, 6972–85.

Gaunt, S. J., Krumlauf, R. and Duboule, D. (1989). Mouse homeo-genes within a subfamily, HOX1.4, -2.6, 5.1, display similar anteroposterior domains of expression on the embryo, but show stage- and tissue-dependent differences in their regulation. *Development*, **107**, 131–41.

Giles, E. and Ambrose, S. H. (1986). Are we all out of Africa? *Nature*, **322**, 21–2.

Gill, P., Jeffreys, A. J. and Werrett, D. J. (1985). Forensic application of DNA fingerprints. *Nature*, **318**, 577–9.

Giuffra, L. A., Lichter, P., Wu, J., Kennedy, J. L., Pakstis, A. J., Rogers, J., Kidd, J. R., Harley, H., Jenkins, T., Ward, D. C. and Kidd, K. K. (1990). Genetic and physical mapping and population studies of a fibronectin receptor beta-subunit-like sequence on human chromosome 19. *Genomics*, **8**, 340–6.

Gonzalez, I. L., Sylvester, J. E., Smith, T. F., Stambolian, D. and Schmickel, R. D. (1990). Ribosomal RNA gene sequences and hominoid phylogeny. *Molecular Biology and Evolution*, **7**, 203–19.

Goodman, M., Koop, B. F., Czelusniak, J., Fitch, D. H. A., Tagle, D. A. and Slightom, J. L. (1989). Molecular phylogeny of the family of apes and humans. *Genome*, **31**, 316–35.

Groot, P. C., Bleeker, M. J., Plonk, J. C., Arwert, F., Mager, W. H., Planta, R. J., Eriksson, A. W. and Frants, R. R. (1989). The human alpha-amylase multigene family consists of haplotypes with variable numbers of genes. *Genomics*, **5**, 29–42.

Hart, C. P., Fainsod, A. and Ruddle, F. H. (1987). Sequence analysis of the murine HOX2.2, -2.3 and -2.4 homeo boxes: Evolutionary and structural comparisons. *Genomics*, **1**, 182–95.

Hartl, D. L. and Clark, A. G. (1989). *Principles of Population Genetics*, 2nd edn. Sunderland, Mass.: Sinauer Associates.

Hill, A. V. S. and Serjeantson, S. W. (ed.) (1989) *The Colonization of the Pacific*. Oxford: Clarendon Press.

Hill, A. V. S., O'Shaughnessy, D. F. and Clegg, J. B. (1989). Haemoglobin and globin gene variants in the Pacific. In *The Colonization of the Pacific*, ed. A. V. S. Hill and S. W. Serjeantson. Oxford: Clarendon Press.

Hixson, J. E., Borenstein, S., Cox, L. A., Rainwater, D. L. and VandeBerg, J. L. (1988a). The baboon gene for apolipoprotein A-I: characterization and identification of DNA polymorphisms for genetic studies of cholesterol metabolism. *Gene*, **74**, 483–90.

Hixson, J. E., Cox, L. and Borenstein, S. (1988b). The baboon apolipoprotein E gene: Structure, expression and linkage with the gene for apolipoprotein C-I. *Genomics*, **2**, 315–23.

Hixson, J. E., Kammerer, C. M., Cox, L. A. and Mott, G. E. (1989). Identification of LDL receptor gene marker associated with altered levels of LDL cholesterol and apolipoprotein B in baboons. *Arteriosclerosis*, **9**, 829–35.

Holland, P. W. H. and Hogan, B. L. M. (1988). Expression of homeo box genes during mouse development: A review. *Genes and Development*, **2**, 773–82.

Human Gene Mapping 10 (1989). Tenth International Workshop on Human Gene Mapping. *Cytogenetics and Cell Genetics*, **51**, 1–1148.

Human Gene Mapping 10.5 (1990). Update to the Tenth International Workshop on Human Gene Mapping. *Cytogenetics and Cell Genetics*, **55**, 1–785.

Innis, M. A., Gelfand, D. H., Sninsky, J. J. and White, T. J. (ed.) (1990). *PCR Protocols: A Guide to Methods and Applications*. San Diego: Academic Press, Inc.

Jeffreys, A. J. (1979). DNA sequence variants in the G-gamma, A-gamma, delta and beta-globin genes of man. *Cell*, **18**, 1–10.

Jeffreys, A., Barrie, P., Harris, S., Fawcett, D., Nugent, Z. and Boyd, C. (1982). Isolation and sequence analysis of a hybrid delta-globin pseudogene from the brown lemur. *Journal of Molecular Biology*, **156**, 487–503.

Jeffreys, A. J., Neumann, R. and Wilson, V. (1990). Repeat unit sequence variation in minisatellites: A novel source of DNA polymorphism for studying variation and mutation by single molecule analysis. *Cell*, **60**, 473–85.

Jeffreys, A. J., Royle, N. J., Wilson, V. and Wong, Z. (1988). Spontaneous mutation rates to new length alleles at tandem-repetitive hypervariable loci in human DNA. *Nature*, **332**, 278–81.

Jeffreys, A. J., Wilson, V. and Thein, S. L. (1985). Hypervariable 'minisatellite' regions in human DNA. *Nature*, **314**, 67–73.

Jones, J. S. and Rouhani, S. (1986). How small was the bottleneck? *Nature*, **319**, 449–50.

Jorde, L. B. (1980). The genetic structure of subdivided human populations: A review. In *Current Developments in Anthropological Genetics. Volume 1: Theory and Methods*, ed. J. H. Mielke and M. H. Crawford. New York: Plenum Press.

Kan Y. W. and Dozy, A. M. (1980). Evolution of the hemoglobin S and C genes in world populations. *Science*, **209**, 388–91.

Kessel, M., Balling, R. and Gruss, P. (1990). Variations of cervical vertebrae after expression of a Hox1.1 transgene in mice. *Cell*, **61**, 301–8.

Kidd, J. R. and Kidd, K. K. (1990). Characterization of the R. Surui and Karitiana at 21 polymorphic DNA loci. *American Journal of Physical Anthropology*, **81**, 249.

Kidd, K. K. (1987). Population genetics of a disease. *Nature*, **327**, 282–3.

Kidd, K. K., Bowcock, A. M., Schmidtke, J., Track, R. K., Ricciuti, F., Hutchings, G., Bale, A., Pearson, P. and Willard, H. F. (1989). Report of the DNA committee and catalogs of cloned and mapped genes and DNA polymorphisms. Human Gene Mapping 10 (1989): Tenth International Workshop on Human Gene Mapping. *Cytogenetics and Cell Genetics*, **51**, 622–947.

Koop, B. F., Goodman, M., Xu, P., Chan, K. and Slightom, J. L. (1986). Primate eta-globin DNA sequences and man's place among the great apes. *Nature*, **319**, 234–8.

Koop, B. F., Tagle, D. A., Goodman, M. and Slightom, J. L. (1989). A molecular view of primate phylogeny and important systematic and evolutionary questions. *Molecular Biology and Evolution*, **6**, 580–612.

Kurst, A., Kastner, P., Petkovich, M., Zelent, A. and Chambon, P. (1989). A third human retinoic acid receptor, hRAR-gamma. *Proceedings of the National Academy of Sciences USA*, **86**, 5310–14.

Lander, E. S. and Botstein, D. (1989). Mapping Mendelian factors underlying quantitative traits using RFLP linkage maps. *Genetics*, **121**, 185–99.

Lerman, L. S., Fischer, S. G., Hurley, I., Silverstein, K. and Lumelsky, N. (1984). Sequence-determined DNA separations. *Annual Review of Biophysics and Bioengineering*, **13**, 399–423.

Lewin, B. (1990). *Genes IV*. New York: Oxford University Press.

Li, W.-H., Wu, C.-I. and Luo, C.-C. (1984). Nonrandomness of point mutation as reflected in nucleotide substitutions in pseudogenes and its evolutionary implications. *Journal of Molecular Evolution*, **21**, 58–71.

Litt, M. and Luty, J. A. (1989). A hypervariable microsatellite revealed by in vitro amplification of a dinucleotide repeat within the cardiac muscle actin gene. *American Journal of Human Genetics*, **44**, 397–401

Long, J. C., Chakravarti, A., Boehm, C. D., Antonarakis, S. and Kazazian, H. (1990). Phylogeny of human beta-globin haplotypes and its implications for human evolution. *American Journal of Physical Anthropology*, **81**, 113–30.

Lu, W.-M., Huang, C.-H., Socha, W. W. and Blumenfeld, O. O. (1990). Polymorphism and gross structure of glycophorin genes in common chimpanzees. *Biochemical Genetics*, **28**, 399–413.

MacCluer, J. W., Kammerer, C. M., Blangero, J., Dyke, B., Mott, G. E., VandeBerg, J. L. and McGill, H. C., Jr (1988). Pedigree analysis of HDL cholesterol concentration in baboons on two diets. *American Journal of Human Genetics*, **43**, 401–13.

MacIntyre, R. J. (ed.) (1985). *Molecular Evolutionary Genetics*. New York: Plenum Press.

Maeda, N., Bliska, J. B. and Smithies, O. (1983). Recombination and balanced chromosome polymorphism suggested by DNA sequences 5' to the human delta-globin gene. *Proceedings of the National Academy of Sciences USA*, **80**, 5012–16.

Maeda, N. and Smithies, O. (1986). The evolution of multigene families: Human haptoglobin genes. *Annual Review of Genetics*, **20**, 81–108.

Maeda, N., Wu, C.-I., Bliska, J. and Reneke, J. (1988). Molecular evolution of intergenic DNA in higher primates: Pattern of DNA changes, molecular clock and evolution of repetitive sequences. *Molecular Biology and Evolution*, **5**, 1–20.

Marks, J. (1983). Hominoid cytogenetics and evolution. *Yearbook of Physical Anthropology*, **26**, 131–59.

Marks, J., Schmid, C. W. and Sarich, V. M. (1988). DNA hybridization as a guide to phylogeny: relations of the Hominoidea. *Journal of Human Evolution*, **17**, 769–86.

Marks, J., Shaw, J.-P., Perez-Stable, C., Hu, W.-S., Ayres, T. M., Shen, C. and Shen, C.-K. J. (1986). The primate alpha-globin gene family: A paradigm of the fluid genome. *Cold Spring Harbor Symposia on Quantitative Biology*, **51**, 499–508.

Matera, A. G., Hellmann, U. and Schmid, C. W. (1990). A transpositionally and transcriptionally competent Alu subfamily. *Molecular and Cellular Biology*, **10**, 5424–32.

Maynard Smith, J. (1990). The Y of human relationships. *Nature*, **344**, 591–2.

Miyamoto, M. M., Slightom, J. L. and Goodman, M. (1987). Phylogenetic relations of humans and African apes from DNA sequences in the pseudo-eta globin region. *Science*, **238**, 369–73.

Mourant, A. E., Kopeć, A. C. and Domanienska-Sobczak, K. (1976). *The Distribution of the Human Blood Groups and other Polymorphisms*. London: Oxford University Press.

Murray, J. C., Mills, K. A., Demopoulos, C. M., Hornung, S. and Motulsky, A. G. (1984). Linkage disequilibrium and evolutionary relationships of DNA variants (restriction fragment length polymorphisms) at the serum albumin locus. *Proceedings of the National Academy of Sciences USA*, **81**, 3486–90.

Myers, R. M., Lumelsky, N., Lerman, L. S. and Maniatis, T. (1985). Detection of single base substitutions in total genomic DNA. *Nature*, **313**, 495–8.

Nakamura, Y., Carlson, M., Krapcho, K., Kanamori, M. and White, R. (1988). New approach for isolation of VNTR markers. *American Journal of Human Genetics*, **43**, 854–9.

Nakamura, Y., Julier, C., Wolff, R., Holm, T., O'Connell, P., Leppert, M. and White, R. (1987a). Characterization of a human 'midisatellite' sequence. *Nucleic Acids Research*, **15**, 2537–47.

Nakamura, Y., Leppert, M., O'Connell, P., Wolff, R., Holm, T., Culver, M., Martin, C., Fujimoto, E., Hoff, M., Kumlin, E. and White, R. (1987b). Variable number of tandem repeat (VNTR) markers for human gene mapping. *Science*, **235**, 1616–22.

Nathans, J., Thomas, D. and Hogness, D. S. (1986). Molecular genetics of human color vision: The genes encoding blue, green and red pigments. *Science*, **232**, 193–202.

Nei, M. (1986). Stochastic errors in DNA evolution and molecular phylogeny. In *Evolutionary Perspectives and the New Genetics*, ed. H. Gershowitz, D. L. Rucknagel and R. E.Tashian. New York: Alan R. Liss.

Nei, M. (1987). *Molecular Evolutionary Genetics*. New York: Columbia University Press.

Nei, M. and Roychoudhury, A. (1982). Genetic relationship and evolution of human races. *Evolutionary Biology*, **14**, 1–59.

Nurse, G. T., Weiner, J. S., and Jenkins, T. (1985). *The People of Southern Africa and their Affinities*. Oxford: Oxford University Press.

Pagnier, J., Mears, J. G., Dunda-Belkhodja, O., Schaefer-Rego, K. E., Beldjord, C., Nagel, R. L. and Labie, D. (1984). Evidence for the multicentric origin of the sickle cell hemoglobin gene in Africa. *Proceedings of the National Academy of Sciences USA*, **81**, 1771–3.

Pamilo, P. and Nei, M. (1988). Relationships between gene trees and species trees. *Molecular Biology and Evolution*, **5**, 568–83.

Preuschoft, H., Chivers, D. J., Brockelman, W. Y. and Creel, N. (ed.) (1984). *The Lesser Apes – Evolutionary and Behavioral Biology*. Edinburgh: Edinburgh University Press.

Ramsay, M. and Jenkins, T. (1987). Globin gene-associated restriction fragment length polymorphisms in Southern African peoples. *American Journal of Human Genetics*, **41**, 1132–44.

Ramsay, M. and Jenkins, T. (1988). Alpha-globin gene cluster haplotypes in the Kalahari San and Southern African Bantu-speaking blacks. *American Journal of Human Genetics*, **43**, 527–33.

Rogers, J. (1989). *Genetic structure and microevolution in a population of Tanzanian yellow baboons*. PhD thesis, Yale University.

Rogers, J., Ruano, G. and Kidd, K. K. (1991). Variability in nuclear DNA among non-human primates: Application of molecular genetic techniques to intra- and inter-species genetic analyses. *American Journal of Primatology* (in press).

Ruano, G., Gray, M. R., Miki, T., Ferguson-Smith, A. C., Ruddle, F. H. and Kidd, K. K. (1990). Monomorphism in humans and sequence differences among higher primates for a sequence tagged site (STS) in homeobox cluster 2 as assayed by denaturing gradient electrophoresis. *Nucleic Acids Research*, **18**, 1314.

Samuelson, L. C., Wiebauer, K., Snow, C. M. and Meisler, M. H. (1990). Retroviral and pseudogene insertion sites reveal the lineage of human salivary and pancreatic amylase genes from a single gene during primate evolution. *Molecular and Cellular Biology*, **10**, 2513–20.

Sarich, V. M. and Wilson, A. C. (1967). Immunological time scale for hominid evolution. *Science*, **158**, 1200–3.

Sarich, V. M., Schmid, C. W. and Marks, J. (1989). DNA hybridization as a guide to phylogeny: a critical analysis. *Cladistics*, **5**, 3–32.

Savatier, P., Trabuchet, G., Chebloune, Y., Faure, C., Verdier, G. and Nigon, V. M. (1987). Nucleotide sequence of the beta-globin genes in gorilla and macaque: The origin of nucleotide polymorphisms in human. *Journal of Molecular Evolution*, **24**, 309–18.

Schmid, C. W. and Marks, J. (1990). DNA hybridization as a guide to phylogeny: Chemical and physical limits. *Journal of Molecular Evolution*, **30**, 237–46.

Scriver, C. R. and Clow, C. L. (1980). Phenylketonuria and other phenylalanine hydroxylation mutants in man. *Annual Review of Genetics*, **14**, 179–202.

Seuanez, H. N., Evans, H. J., Martin, D. E. and Fletcher, J. (1979). Inversion of chromosome-2 that distinguishes between Bornean and Sumatran orangutans. *Cytogenetics and Cell Genetics*, **23**, 137–40.

Slatkin, M. and Maddison, W. P. (1989). A cladistic measure of gene flow inferred from the phylogenies of alleles. *Genetics*, **123**, 603–13.

(1990). Detecting isolation-by-distance using phylogenies of genes. *Genetics*, **126**, 249–60.

Smouse, P. E. and Li, W.-H. (1987). Likelihood analysis of mitochondrial restriction cleavage patterns for the human–chimpanzee–gorilla trichotomy. *Evolution*, **41**, 1162–76.

Sokal, R. R. (1988). Genetic, geographic and linguistic distances in Europe. *Proceedings of the National Academy of Sciences USA*, **85**, 1722–6.

Stanyon, R., Chiarelli, B., Gottlieb, K. and Patton, W. H. (1986). The phylogenetic and taxonomic status of *Pan paniscus*: A chromosomal perspective. *American Journal of Physical Anthropology*, **69**, 489–98.

Suarez, B. K., Crouse, J. D. and O'Rourke, D. H. (1985). Genetic variation in North Amerindian populations: The geography of gene frequencies. *American Journal of Physical Anthropology*, **67**, 217–32.

Tagle, D. A., Koop, B. F., Goodman, M., Slightom, J. L., Hess, D. L. and Jones, R. T. (1988). Embryonic epsilon and gamma globin genes of a prosimian primate (*Galago crassicaudatus*): nucleotide and amino acid sequences, developmental regulation and phylogenetic footprints. *Journal of Molecular Biology*, **203**, 439–55.

Trabuchet, G., Chebloune, Y., Savatier, P., Lachuer, J., Verdier, G. and Nigon, V. M. (1987). Recent insertion of an Alu sequence in the beta-globin gene cluster of the gorilla. *Journal of Molecular Evolution*, **25**, 288–91.

Tsujibo, H., Tiano, H. F. and Li, S. (1985). Nucleotide sequences of the cDNA and an intronless pseudogene for human lactate dehydrogenase-A isozyme. *European Journal of Biochemistry*, **147**, 9–15.

Vanin, E. F. (1985). Processed pseudogenes: characteristics and evolution. *Annual Review of Genetics*, **19**, 253–72.

Verga, V., Dos Santos, M., Jenkins, T., Marques, I., Povey, S. and Ramsey, M. (1989). An arbitrary single copy DNA sequence VC85 (D1S85) detects a 500 basepair insertion/deletion polymorphism on chromosome 1. *Nucleic Acids Research*, **17**, 4007.

Vigilant, L., Pennington, R., Harpending, H., Kocher, T. D. and Wilson, A. C. (1989). Mitochondrial DNA sequences in single hairs from a southern African population. *Proceedings of the National Academy of Sciences USA*, **86**, 9350–4.

Wainscoat, J. S., Hill, A. V. S., Boyce, A. L., Flint, J., Hernandez, M., Thein, S. L. Old, J. M., Lynch, J. R., Falusi, A. G., Weatherall, D. J. and Clegg, J. B. (1986). Evolutionary relationships of human populations from an analysis of nuclear DNA polymorphisms. *Nature*, **319**, 491–3.

Weber, J. L. and May, P. E. (1989). Abundant class of human DNA polymorphisms which can be typed using the polymerase chain reaction. *American Journal of Human Genetics*, **44**, 388–96.

Weiner, A. M., Deininger, P. L. and Efstradiatis, A. (1986). Nonviral retroposons: genes, pseudogenes and transposable elements generated by the reverse flow of genetic information. *Annual Review of Biochemistry*, **55**, 631–61.

Weiss, K. M. (1988). In search of times past: Gene flow and invasion in the generation of human diversity. In *Biological Aspects of Human Migration*, ed. G. C. N. Mascie-Taylor and G. W. Lasker. Cambridge: Cambridge University Press.

(1989). A survey of human biodemography. *Journal of Quantitative Anthropology*, **1**, 79–151.

Weiss, K. M., Ferrell, R. E. and Hanis, C. L. (1984). A New World Syndrome of metabolic diseases with a genetic and environmental basis. *Yearbook of Physical Anthropology*, **27**, 153–78.

Weiss, M. L. (1987). Nucleic acid evidence bearing on hominoid relationships. *Yearbook of Physical Anthropology*, **30**, 41–73.

Weiss, M. L., Wilson, V., Chan, C., Turner, T. and Jeffreys, A. J. (1988). Application of DNA fingerprinting to Old World Monkeys. *American Journal of Primatology*, **16**, 73–9.

Wilkie, A. O. M., Higgs, D. R., Rack, K. A., Buckle, V. J., Spurr, N. K., Fischel-Ghodsian, N., Ceccherini, I., Brown, W. R. A. and Harris, P. C. (1991). Stable length polymorphism of up to 260kb at the tip of the short arm of human chromosome 16. *Cell*, **64**, 595–606.

Willard, H. F., Waye, J. S., Skolnick, M. H., Schwartz, C. E., Powers, V. E. and England, S. B. (1986). Detection of restriction fragment length polymorphisms at the centromeres of human chromosomes by using chromosome-specific alpha-satellite DNA probes: Implications for development of centromere-based genetic linkage maps. *Proceedings of the National Academy of Sciences USA*, **83**, 5611–5.

Williams, R. C. (1989). Restriction fragment length polymorphism (RFLP). *Yearbook of Physical Anthropology*, **32**, 159–84.

Wong, C., Dowling, C. E., Saiki, R. K., Higuchi, R. G., Erlich, H. A. and Kazazian, H. H., Jr (1987). Characterization of beta-thalassemia mutations using direct sequencing of amplified single copy DNA. *Nature*, **330**, 384–6.

Wong, Z., Royle, N. J. and Jeffreys, A. J. (1990). A novel human DNA polymorphism resulting from transfer of DNA from chromosome 6 to chromosome 16. *Genomics*, **7**, 222–34.

Wyman, A. and White, R. (1980). A highly polymorphic locus in human DNA. *Proceedings of the National Academy of Sciences USA*, **77**, 6754–8.

Yunis, J. J and Prakash, O. (1982). The origin of man: A chromosomal pictorial legacy. *Science*, **208**, 1145–8.

Zimmer, E. A., Martin, S. L., Beverly, S. M., Kan, Y. W. and Wilson, A. C. (1980). Rapid duplication and loss of genes coding for the alpha globin chains of hemoglobin. *Proceedings of the National Academy of Sciences USA*, **77**, 2158–62.

3 Molecular genetics of atherosclerosis in human and non-human primates

JAMES E. HIXSON

Introduction

The motivation for studies dissecting genetic and environmental factors that influence complex traits is especially strong, since such traits include common diseases with large effects on mortality in industrialized human societies. Common diseases (e.g. diabetes, hypertension and atherosclerosis) aggregate in families, but are not transmitted by simple Mendelian inheritance. The lack of simple Mendelian inheritance may stem from many causes, including (1) the interaction of many genes to affect a single phenotype (epistasis), (2) the action of a single gene to affect several traits (pleiotropy), (3) different expression of identical genes in different individuals (expressivity and penetrance) and (4) interaction between genes and environmental factors such as nutrition.

With the advent of recombinant DNA technology, new methods to examine directly gene structure, regulation and DNA polymorphisms are being used in family studies to characterize single gene defects that cause diseases such as cystic fibrosis, muscular dystrophy, retinoblastoma and Huntington's disease (for review see Caskey, 1987). But genetic studies of common diseases with complex patterns of inheritance will also profit from application of molecular methods. General approaches that are useful for these include (1) family or population studies that relate disease traits to RFLPs in candidate genes, those loci encoding proteins involved in related physiological processes, or (2) family studies to detect coseg-regation of DNA variants at unrelated loci (random markers) that mark known chromosomal locations. The second approach will be increasingly useful for studies of common diseases, as DNA probes, RFLPs and other DNA polymorphisms are identified that span the chromosomal map of the human genome.

This chapter describes recent molecular genetic studies of athero-sclerosis, a disease resulting from lifelong development of arterial lesions that ultimately block vascular flow and result in myocardial infarct (blockage to heart) or stroke (blockage to brain). The analytical and methodological approaches that are used for genetic studies of athero-sclerosis provide a useful model for studies of other common diseases that result from complex interactions between genes and environmental factors.

Candidate genes of atherosclerosis

Risk factors of atherosclerosis

While the molecular basis of atherosclerosis is not yet known, epidemio-logical studies show that serum levels of cholesterol and lipoproteins (particles that carry cholesterol) are risk factors for atherosclerosis and subsequent heart disease. In particular, elevated levels of circulating low density lipoproteins (LDL) are associated with increased risk of athero-sclerosis, while elevated levels of high density lipoproteins (HDL) are associated with decreased risk. Serum levels of cholesterol and lipopro-teins for a particular individual may be affected by many genes in the pathway of cholesterol metabolism, as well as by environmental factors such as dietary intake of saturated fat and cholesterol. The candidate genes for atherosclerosis are those that affect cholesterol and lipoprotein levels, including genes for apolipoproteins (proteins associated with lipoprotein particles), lipoprotein receptors, and cholesterol processing enzymes.

Lipoproteins carry cholesterol in the vascular system by packaging hydrophobic cholesteryl esters in a core surrounded by phospholipids. Lipoproteins also contain proteins (apolipoproteins) that bind choles-terol for solubilization and that interact with cellular receptors to deliver lipoprotein particles to target tissues. Specific size classes of lipoproteins contain different apolipoproteins (for review see Breslow, 1985). For example, apolipoprotein B (apo B) is found on LDL particles and binds with high affinity to LDL receptors for uptake and catabolism of LDL by liver cells. Apo AI and apo AII are found on HDL particles. Apo E is found on several different lipoproteins and binds two different cellular receptors (LDL receptor and apo E receptor). Many genes encoding apolipoproteins have been cloned, sequenced and mapped to human chromosomes. Many of the apolipoprotein genes are distributed in clusters on chromosome 11 (apo AI, CIII and AIV, Bruns *et al.*, 1984) and on chromosome 19 (apo CII, E, and CI: Myklebost and Rogne, 1988). These cloned apolipoprotein genes have been used as probes to identify RFLPs that distinguish particular alleles for population studies (for review see Humphries, 1988).

Lipoprotein receptors

Lipoproteins are delivered to target cells by cellular receptors that bind apolipoprotein moieties of lipoprotein particles. The surfaces of liver cells contain LDL receptors that bind lipoproteins containing apo B or apo E. After binding, the LDL receptor–lipoprotein complex enters the liver cell, the lipoprotein is degraded to release free cholesterol and amino acids, and the LDL receptor returns to the membrane to continue the cycle. The human LDL receptor gene has been cloned. It is a large gene containing 18 exons (sequences encoding protein) separated by 17 introns (non-coding sequences) (Sudhof *et al.*, 1985). These exons encode several domains with different functions including a region for lipoprotein binding, a trans-membrane region to anchor the LDL receptor, and a cytoplasmic domain that localizes the receptor in coated pits on hepatic membranes (for review see Brown and Goldstein, 1986).

Hepatic cells also possess another receptor that binds apo E moieties on lipoprotein particles. This apo E receptor is now known as the LDL receptor-related protein (LRP) because of structural homologies with the LDL receptor (Herz *et al.*, 1988). The LRP is much larger than the LDL receptor (4544 versus 839 amino acids), and is encoded by a gene located on human chromosome 12 (Myklebost *et al.*, 1989). This receptor binds chylomicron remnants and VLDL particles that bear apo E for uptake and catabolism by hepatic cells (Kowal *et al.*, 1989).

Cholesterol processing enzymes

Cholesterol metabolism also requires the activities of several serum enzymes that process cholesterol and lipoproteins for packaging and transport. Lecithin–cholesterol acyltransferase (LCAT) is a serum enzyme that catalyses the esterification of free cholesterol for packaging in hydrophobic cores of HDL particles (for review see Small, 1987). The apo AI and apo AII components of HDL also participate in packaging by stimulation of LCAT activities. After LCAT packages cholesterol in HDL particles, cholesteryl ester transfer protein (CETP) catalyses the transfer of cholesteryl esters from HDL particles to LDL and VLDL particles for removal by liver cells (for review see Tall, 1986). These activities of LCAT and CETP are part of 'reverse cholesterol transport', the transport of excess cholesterol from peripheral tissues for removal by hepatic cells (for review see Reichl and Miller, 1989). Lipoprotein lipase (LPL) is another serum enzyme in the pathway of cholesterol metabolism. LPL catalyses the hydrolysis of fatty acids in lipoprotein particles to produce free fatty acids for peripheral tissues (for review see Small, 1987). These LPL-processed lipoproteins are then removed by hepatic receptors, or further processed to form LDL or HDL.

The baboon model for genetic studies of atherosclerosis

Controlling genetic and dietary factors for studies of atherosclerosis in baboons

Genetic studies of atherosclerosis in human subjects have been particularly difficult because of effects caused by dietary factors. Baboons are being used in studies of atherosclerosis because their diets can be carefully designed and maintained. Like humans, baboons react with a modest increase of cholesterol levels when fed atherogenic diets (high cholesterol and saturated fat) (McGill *et al.*, 1981a). In addition, baboons develop atherosclerotic lesions in their natural habitats, a process that is further stimulated by atherogenic diets in captivity (McGill *et al.*, 1981b).

Dietary effects are examined in baboons by comparing risk factors (serum levels of cholesterol, lipoproteins, and apolipoproteins) after long-term feeding on basal diets (10% of kcal as fat, 0.03 mg/kcal cholesterol) with risk factors after feeding for 7 weeks on an atherogenic diet (40% of kcal as fat, 1.7 mg/kcal cholesterol). This approach not only permits detection of genes that affect cholesterol metabolism under different diets, but also allows characterization of genotypic response to dietary intake of high levels of cholesterol and saturated fats similar to human diets. In addition to controlling dietary factors, genetic factors can be controlled by breeding programmes to examine particular risk factors of atherosclerosis. For example, 23 sire lineages have been established that carry genes for a variety of lipoprotein phenotypes (MacCluer *et al.*, 1988).

Complex segregation analysis of HDL levels in baboons

The first step in genetic studies in baboons of known pedigree is the determination of relative contributions of genetic and non-genetic factors on particular phenotypes of cholesterol metabolism. The method, called complex segregation analysis, tests different hypotheses that account for the observed distribution of phenotypes among family members (Mac-Cluer *et al.*, 1987). Effects of genetic and non-genetic factors are tested by comparing models that include different assumptions about relative contributions of single genes with large effects (major genes), many genes with small effects (polygenes), and random non-genetic effects. The mode of major gene transmission is determined by comparing models that assume recessive, dominant, or codominant inheritance.

Complex segregation analyses of HDL levels in baboons of known pedigree have detected effects of major genes, polygenes, and random factors on basal and atherogenic diets (MacCluer *et al.*, 1988). The best models for baboons on both diets included codominant expression of a major gene with a rare allele for high levels of HDL. In addition to HDL levels, comparison of diets allowed analysis of response to high cholesterol, saturated fat diets. Unlike results from genetic studies of mice (Ath–1: Paigen *et al.*, 1987), complex segregation analyses did not detect major gene effects, but did detect effects of polygenic factors on changes of HDL levels in response to atherogenic diets.

Methods for molecular genetic studies in baboons

While quantitative segregation analysis detects the effects of major genes and polygenes, the identity of these genes is not determined. In genetic studies of atherosclerosis in baboons, the 'candidate gene' approach is being used to identify the major genes and polygenes that determine lipoprotein phenotypes. Figure 3.1 illustrates methods for molecular genetic studies to detect RFLPs in candidate genes.

Extraction of white cell DNA from blood samples

Molecular genetic studies rely on genomic DNA that is purified from a particular individual. Genomic DNA for population studies is usually extracted from white cells (step 1, Fig. 3.1) that are isolated from whole blood samples. In baboon studies, white cells are separated from whole blood samples on gradients (Hixson *et al.*, 1988a). The white cells can be frozen for storage, or used directly for DNA extraction. For DNA extraction (step 2, Fig. 3.1), baboon white cells are incubated with detergents to lyse the cells, and treated with proteolytic enzymes to degrade cellular proteins (Hixson *et al.*, 1988a). After incubation, the mixture becomes highly viscous because of the nuclear DNA released into solution. The DNA is further purified from cellular proteins by extraction with organic reagents. This extraction requires mixing of the

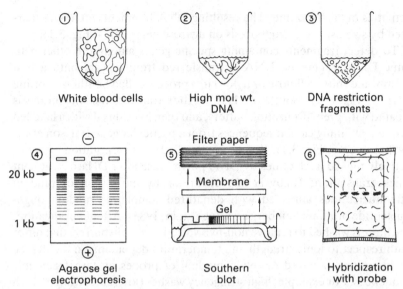

Fig. 3.1. Overview of DNA extraction from white cells and Southern blotting.
Steps 1 and 2 show white cells that are extracted for preparation of high
molecular weight DNA (long curved lines). Step 3 shows treatment of DNA
with restriction enzymes (circles) to produce DNA fragments (short curved
lines). Step 4 shows agarose gel electrophoresis (anode shown by +, cathode
by −) of DNA fragments of various sizes (20 kb marks large fragments that
migrate the least distance, 1 kb marks small fragments that migrate the farthest
distance). Step 5 shows Southern blotting to transfer DNA fragments from gel
to membrane (arrows mark capillary flow of buffer that is drawn to filter
papers). Step 6 shows hybridization of the DNA laden membrane with
radiolabelled probes (shown by short curved lines).

DNA solution with organic reagents and centrifugation to separate the
organic phase from the aqueous (DNA-containing) phase. After several
extractions, the DNA is precipitated by adding salt (e.g. sodium chloride
or sodium acetate) and ethanol. The precipitated DNA is harvested by
spooling on glass rods or by centrifugation, then dissolved in buffers
containing EDTA to inhibit nucleases that degrade DNA.

Southern blotting of white cell DNA

Most DNA polymorphisms (RFLPs) are detected because of the alter-
ations in specific sequences that are sites for cutting by bacterial enzymes
(restriction enzymes). To cut white cell DNA (step 3, Fig. 3.1), DNA is
incubated with a particular restriction enzyme in buffers that optimize

activities of each enzyme. The resulting DNA fragments are then separated by size using electrophoresis on agarose gels (step 4, Fig. 3.1).

To detect fragments containing specific genes among the other genomic DNA fragments, DNA is transferred from the gel onto a thin membrane (nitrocellulose or nylon) in a process called Southern blotting (step 5, Fig. 3.1) (Southern, 1975). After transfer, the membrane is treated with prehybridization buffers, and then hybridized with a labelled probe containing cloned sequences from a particular gene (Hixson *et al.*, 1988a) (step 6, Fig. 3.1). Probes are labelled by incorporation of radioactively tagged nucleotides by DNA polymerase after DNase treatment (nick translation; Rigby *et al.*, 1977), or by elongation of random oligonucleotides annealed with denatured templates (random oligonucleotide priming: Feinberg and Vogelstein, 1983). After hybridization, the filter is washed to remove non-hybridized probes in buffers that differ with respect to ionic strength and temperature depending on the degree of homology required to retain binding of probes by target genomic sequences. For example, high stringency washes (low ionic strength, high temperature) are used in probing human genes with identical human gene probes, but lower stringencies (high ionic strength, low temperature) would be used in detecting genes from a different species (i.e. mouse) with human probes.

Determining genotypes for a hypothetical RFLP using Southern blots

Allelic sequence differences can be detected by abolishing or creating cleavage sites for any of a variety of different restriction enzymes. In the following example, the restriction enzyme Eco RI is used to cut DNA at the nucleotide sequence GAATTC. Eco RI can distinguish an allele that contains a substitution in this sequence (e.g. GTATTC) because DNA for that allele will not be cut by Eco RI. Fig. 3.2 shows how Eco RI is used to detect A versus B alleles at a hypothetical gene locus by Southern blotting with radiolabelled probes. On the left is a gene map that shows Eco RI sites in DNA from the A allele separated by 10 kb, and from the B allele which contains in addition a sequence substitution that creates a new Eco RI site (polymorphic site). Above the maps, a hypothetical probe is shown that spans the region containing the polymorphism. On the right is a hypothetical Southern blot that shows DNA from each genotype (AA, AB, BB) cut with Eco RI, followed by electrophoresis on an agarose gel, and hybridized with the gene probe. The lane marked AA contains a single 10 kb fragment corresponding to two copies of the A

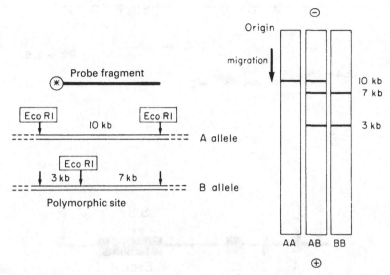

Fig. 3.2. Gene map and Southern blot to distinguish alleles for an Eco RI RFLP in a hypothetical gene. Left, a gene map showing Eco RI sites in A and B alleles, and a probe fragment that hybridizes to Eco RI fragments. Right, a Southern blot showing Eco RI-digested DNAs from each genotype (AA, AB, BB) after hybridization with the probe. The sizes of hypothetical fragments are shown (in kb) to the right of the gel.

allele, the lane marked BB contains 7 and 3 kb fragments corresponding to the presence of the polymorphic Eco RI site in two copies of the B allele, and the lane marked AB contains 10, 7 and 3 kb fragments corresponding to both A and B alleles.

Molecular genetic studies of lipoprotein phenotypes in baboons

Identification of DNA polymorphisms in candidate genes

To identify those restriction enzymes that detect RFLPs, white cell DNA from 20 baboons is cut with many different restriction enzymes. The cut DNA fragments are separated by electrophoresis, transferred to membranes, and hybridized with cloned probes that are radiolabelled by nick translation (Rigby *et al.*, 1977) or random oligonucleotide priming (Feinberg and Vogelstein, 1983). If no polymorphism is detected, all of the hybridizing fragments will be identical for all baboons. If a particular restriction enzyme detects a polymorphism, different fragments will be observed with sizes corresponding to the presence or absence of cut site(s).

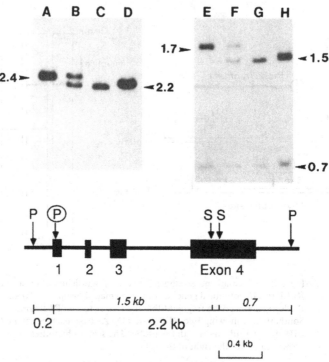

Fig. 3.3. Southern blots of baboon and human apo AI RFLP genotypes, and localization of polymorphic Pst I cleavage sites in baboon DNA. The top left panel (lanes A–D) shows a Southern blot of Pst I-digested baboon DNAs (P1P1 genotype in lane A, P1P2 in lane B, P2P2 in lane C) and human DNA (lane D) after hybridization with a human apo AI cDNA probe. Lanes E–H show a Southern blot of the same DNAs after simultaneous cleavage with Pst I and Sst I to map the baboon Pst I RFLP. The sizes of hybridizing fragments are shown adjacent to the blots (in kb). A map of the Pst I sites (marked P, polymorphic site is circled) and Sst I sites (S) aligned with the human Apo AI gene map (exons are numbered filled boxes) is shown below the Southern blots. Distances between the cleavage sites are given below the map.

Figure 3.3 shows Southern blots of DNA from three baboons representing each genotype for a Pst I RFLP in the apo AI gene (lanes A–C). The location of the polymorphic site (as determined by comparison with Pst I-cut human DNA, lane D; and simultaneous cleavage of baboon and human DNA with Sst I, lanes E–H) is shown below the blots. Homozygotes for the rarer allele possess Pst I sites in the 5′ region of the Apo AI gene resulting in 2.2 kb hybridizing fragments, homozygotes for the common allele lack these sites and produce 2.4 kb fragments, and heterozygotes have both alleles with corresponding 2.2 and 2.4 kb fragments.

Repeating this process with cloned probes for different genes, DNA polymorphisms at many candidate loci have been identified in the baboon genome. In addition to the Pst I RFLP, a polymorphic Pvu II site has been detected in the 3' flanking region of the apo AI gene in baboons (Hixson, 1987). At the LDL receptor gene locus, a polymorphic Ava II site has been detected and mapped to intron 17 (Hixson *et al.*, 1989). RFLPs have also been identified in genes encoding LCAT (Hixson *et al.*, 1990), CETP and LPL.

The Ava II RFLP in LDL receptor genes marks alleles that alter LDL and Apo B levels in baboons

After identification of DNA polymorphisms, RFLP genotypes are determined for large numbers of baboons for statistical analysis to detect genetic effects on lipoprotein phenotypes. For example, after identification of the Ava II RFLP in LDL receptor genes, genotypes were determined for 211 baboons to test for effects on serum levels of LDL and apo B (Hixson *et al.*, 1989). The frequency of homozygotes for the common allele (A1A1) was 0.62, the frequency of heterozygotes (A1A2) was 0.33, and the frequency of homozygotes for the rarer allele (A2A2) was 0.04.

Using a modification of measured genotype analysis, likelihoods were compared for a model which included mean levels of LDL and apo B for each of the three genotypes versus a model in which there were no differences among genotypes (Hixson *et al.*, 1989). This comparison showed that LDL receptor genotypes were significantly associated with different concentrations of LDL and apo B on both diets. A1A1 homozygotes had higher mean LDL levels on basal and atherogenic diets (55.83 and 117.40 mg/dl, respectively) than did A2A2 homozygotes (41.23 and 81.87 mg/dl, respectively), and heterozygotes had intermediate LDL levels (47.55 and 89.98 mg/dl, respectively). Similarly, A1A1 homozygotes had the highest levels of apo B on the atherogenic diet (60.25 mg/dl), A2A2 homozygotes had the lowest levels (47.56 mg/dl), and heterozygotes had intermediate apo B levels (53.13 mg/dl). A1A1 homozygotes also had the highest apo B levels on the basal diet (41.95 mg/dl), but A2A2 homozygotes had apo B levels similar to those of heterozygotes. While the molecular bases of these differences among genotypes are not yet known, the Ava II polymorphism may identify alleles encoding LDL receptors with different properties of apo B binding or internalization, or that differ in cellular distribution or number.

Structure and evolution of baboon apolipoprotein genes

Construction of a baboon genomic library

The molecular characterization of baboon genes of cholesterol metabolism began by isolation of apolipoprotein genes from genomic libraries. To isolate apolipoprotein genes, a genomic library was constructed using a bacteriophage host DNA (vector) (Hixson et al., 1988b). Genomic libraries are constructed by ligation of genomic DNA fragments into a non-essential region of the vector DNA. The ligation products are packaged by enzymes in cell free extracts (called in vitro packaging) to form viable phage particles. The recombinant phage particles are used to infect a host bacterial strain (E. coli) and spread on culture plates to produce a bacterial lawn with cleared regions (plaques) where phage growth causes bacterial lysis. The target gene is identified by hybridization with a radiolabelled probe (as in Southern blots) after transfer of phage DNA to membranes.

The baboon Apo E gene

The baboon apo E gene was cloned by hybridization of the genomic library with a probe containing human apo E coding sequences (Hixson et al., 1988b). The nucleotide sequence of the baboon apo E gene was determined by subcloning fragments from the original clone (9 kb insert) into another vector (bacteriophage M13) for chain-termination sequencing. This method uses nucleotide analogues (dideoxyribonucleotides) that are incorporated, but prevent further elongation of primers with DNA polymerase to form sequence ladders. Nucleotide sequencing showed that the baboon apo E gene contains 4 exons separated by 3 introns.

The baboon and human apo E genes were aligned to examine evolutionary changes since divergence of these primate species (Hixson et al., 1988b). On average, the baboon and human exons were 96% identical, and introns 2 and 3 were 87% identical. Unlike the other introns, the extent of intron 1 divergence was more like coding sequences (95% identical), perhaps attributable to conservation of regulatory sequences. Surprisingly, many of the nucleotide substitutions between human and baboon genes were non-synonymous substitutions resulting in amino acid changes. Fifty per cent of the substitutions were non-synonymous, resulting in a non-synonymous substitution rate approximately 1.6 times that for an average mammalian gene. To examine the effects of amino acid changes caused by non-synonymous substitutions, the hydrophilicity

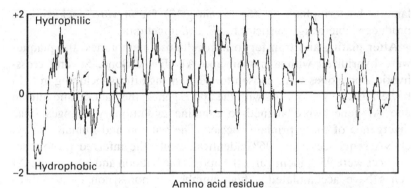

Fig. 3.4. Hydrophilicity profiles of baboon and human apo E derived from nucleotide sequences. The y-axis shows the hydrophilicity value for each amino acid residue of baboon apo E shown on the x-axis (solid line graph). Regions of the human profile that differ from baboon apo E are shown by dotted lines. The eight 22 aa repeats that form amphipathic helices are indicated by thin vertical lines.

profiles (distribution of amino acids with polar versus non-polar side chains) of the deduced baboon and human apo E proteins were compared (Fig. 3.4). This comparison showed that the protein profiles are very similar because of substitutions of amino acids with similar properties. Despite extensive divergence, higher order structures that are critical to apo E function have been evolutionarily conserved.

Using Southern blots, apo CI sequences were also detected on the recombinant clone containing apo E sequences (Hixson *et al.*, 1988b). Nucleotide sequencing showed that these genes were separated by 4.3 kb, similar to the distribution of human apo E and apo CI genes. Apparently, this close linkage arrangement has been conserved since the divergence of baboon and human species.

Isolation and characterization of baboon Apo AI cDNA

In addition to the genomic library, a copy DNA (cDNA) library was constructed from liver RNA for isolation of clones containing apo AI sequences (Hixson *et al.*, 1988a). The library was constructed by elongating synthetic poly(dT) primers on hepatic mRNAs with reverse transcriptase (copies RNA to make DNA). The RNA was then degraded (by RNase), and the newly synthesized DNA strand was elongated to form a double-stranded DNA product. After addition of synthetic linkers that add Eco RI cleavage sites to each end, the cDNA was cloned into a lambda bacteriophage vector (strain gt10) for *in vitro* packaging to produce viable phage particles for *E. coli* infection.

After plating and transfer to nitrocellulose membranes, the plaques were hybridized with a human apo AI cDNA clone. Several cross-hybridizing clones were chosen for sequencing after subcloning in M13 bacteriophage vectors. As with the apo E genes, the baboon and human apo AI genes were aligned to examine evolutionary changes since divergence of these primate species. The baboon and human apo AI cDNA sequences were 96% identical, while the inferred protein sequences were 95% identical. Like apo E, the baboon and human apo AI genes have accumulated a high number of non-synonymous substitutions, but share amino acid sequences with similar hydrophilicity profiles, reflecting evolutionary maintenance of apo AI structure and function.

Molecular genetic studies of atherosclerosis in man

Candidate genes and atherosclerosis in man

Because of the effect of atherosclerosis on human health and mortality, candidate gene studies of atherosclerosis are under way for populations in many different countries. Thus far, most molecular genetic studies have focused on particular diseases such as familial hypercholesterolaemia (FH) with elevated cholesterol levels, hypertriglyceridaemia (elevated triglyceride levels), hypobetalipoproteinaemia (low apo B levels), and familial LPL deficiency (low LPL levels). In most cases, comparisons of disease and normal genes revealed mutations that abolish gene function such as deletions, inversions, or substitutions that introduce premature translational stop codons.

RFLPs mark apolipoprotein gene alleles that affect lipoprotein levels and atherosclerosis

Apolipoprotein gene RFLPs are being used in two types of studies, (1) studies of unrelated individuals to detect associations with risk factors or heart disease (association studies) and (2) studies of related individuals to detect cosegregation of RFLPs with lipoprotein or disease phenotypes (linkage analysis). Many association studies have used RFLPs to search for differences that distinguish patients from control subjects (case-control studies). Frequencies of genotypes are compared between patient and control groups to detect differences that may reflect genotypic effects on disease endpoints. Conversely, subjects may be divided according to

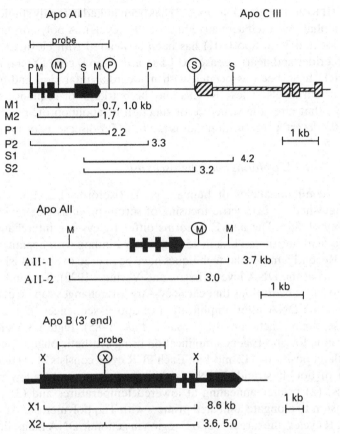

Fig. 3.5. Apolipoprotein gene maps of various RFLPs, and sizes of fragments that distinguish each genotype. The gene maps (apo AI–CIII, apo AII, and apo B) show exon regions (filled boxes or hatched boxes for apo CIII) and their transcriptional orientations (arrow shape), intron regions (thick line, hatched thick line for apo CIII), and intergenic or non-transcribed flanking sequences (thin lines). The locations of cleavage sites are shown above the map (P, Pst I; M, Msp I; S, Sst I; X, Xba I), and the polymorphic sites are circled. Under each gene map are the designations for each RFLP allele, and the location and sizes of corresponding fragments (brackets and sizes in kb).

genotype to compare mean levels of cholesterol, lipoproteins, or apolipo-proteins to detect significant effects from a particular allele.

Figure 3.5 shows maps of several RFLPs (circled) for human genes encoding apo AI-CIII, apo AII, and apo B. Several of these polymorphisms have been used to determine their effects on lipoprotein pheno-types and heart disease (for review see Cooper and Clayton, 1988; Humphries, 1988). For example, a polymorphic Pst I site at the apo

AI-CIII locus (located 3' to apo AI) has been linked in family studies with hypoalphalipoproteinaemia (reduced HDL levels). A polymorphic Sst I site (located 3' to apo CIII) has been associated with elevated serum triglycerides and heart disease. RFLPs for the apo B gene (Xba I, Msp I, Eco RI) have been associated with myocardial infarction and altered lipoprotein levels. These RFLPs may be markers for sequence substitutions that alter the structure or function of apolipoproteins, or may directly detect a substitution that is the basis for disease association.

Apo E isoforms and atherosclerosis

Three common alleles of human apo E (isoforms E2, E3, E4) are distinguished by isoelectric focusing of serum proteins (for review see Mahley, 1988). The apo E isoforms differ by cys–arg interchanges at amino acid positions 112 and 158 that alter binding with hepatic receptors. Recently, molecular techniques have been used to distinguish apo E isoforms at the DNA level (Hixson and Vernier, 1990). The underlying nucleotide substitutions that cause cys–arg interchanges can be detected in genomic DNA after amplification of apo E sequences by the polymerase chain reaction or PCR (Saiki *et al.*, 1985). Figure 3.6 shows a schematic for apo E gene amplification using synthetic oligonucleotides that flank positions 112 and 158. Each PCR cycle consists of (1) denaturation of double-stranded DNA at high temperatures to allow primer access, (2) primer annealing at lowered temperatures and (3) primer extension to elongate annealed primers with Taq polymerase. After 25–30 PCR cycles, the targeted apo E region in genomic DNA is amplified by approximately one million-fold, allowing direct gel analysis of apo E sequences rather than detection by Southern blots. The amplified apo E sequences are then cut with Hha I for electrophoresis on polyacrylamide gels. Hha I distinguishes each isoform by cutting at GCGC that encodes arg at positions 112 (E4 isoform) and 158 (E3 and E4), but does not cut at GTGC that encodes cys at positions 112 (E2 and E3) and 158 (E2).

Studies in many different human populations have detected effects of apo E isoforms on serum levels of cholesterol and lipoproteins, and on heart disease (for review, see Davignon *et al.*, 1988). The E2 isoform is associated with hypertriglyceridaemia and lowered cholesterol levels, most likely owing to reduced receptor binding relative to the common E3 isoform (for review see Mahley, 1988). Because the E2 isoform binds poorly for uptake by apo E and LDL receptors, cholesterol levels may be reduced in hepatic cells, resulting in stimulation of LDL receptor production and increased clearance of serum cholesterol.

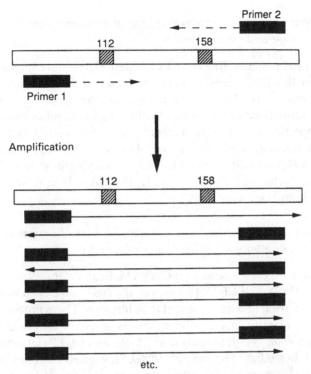

Fig. 3.6. Amplification of apo E sequences in genomic DNA by the polymerase chain reaction (PCR). At the top, a map of apo E is shown that contains amino acid positions 112 and 158 (numbered hatched boxes) that distinguish apo E isoforms (E2, E3, E4). The location of flanking primers (filled boxes numbered 1 and 2) and direction of extension by polymerase (dotted arrows) are shown adjacent to the map. After 20–30 cycles of amplification (downward arrow), the primers have been extended (solid arrows) to produce a million-fold amplification of the apo E sequences for determination of apo E isoform genotypes.

A recent study showed that apo E isoforms also affect development of atherosclerosis (Hixson *et al.*, 1991). In a study of autopsied young persons (15–34 years of age), arterial surfaces were measured for atherosclerotic lesions and liver DNA samples were used to type apo E isoforms. Among the common genotypes (E2E3, E3E3, E3E4), E2E3 heterozygotes had the fewest lesions and E3E4 heterozygotes had the most lesions. Like other population studies, the E2 allele was associated with lower cholesterol levels and the E4 allele with higher levels. These effects on cholesterol levels may explain the associations of apo E isoforms with differences in amounts of atherosclerotic lesions.

Effects of LDL receptor gene variation on lipoprotein phenotype and atherosclerosis

In studies of FH patients, Brown and Goldstein (1986) have shown that mutations in the LDL receptor gene can cause dramatic changes in cholesterol metabolism that result in premature atherosclerosis. Several different genetic defects have been identified in FH families that alter various properties of the LDL receptor (e.g. synthesis, processing, transport, distribution, or ligand binding). In recent efforts to find genetic markers for FH, many different RFLPs have been identified in the LDL receptor gene including RFLPs for Pvu II, Pst I, Ava II, and Nco I (for review see Humphries *et al.*, 1989).

Effects of variation in genes encoding cholesterol processing enzymes on lipoprotein levels and atherosclerosis

In recent years, clones encoding LCAT (McLean *et al.*, 1986), LPL (Wion *et al.*, 1987), and CETP (Drayna *et al.*, 1987) have been isolated and sequenced from human hepatic cDNA libraries. These probes have been used to identify DNA polymorphisms in genes encoding LPL (Chamberlain *et al.*, 1989; Gotoda *et al.*, 1989) and CETP (Drayna and Lawn, 1987). In studies of human subjects, LPL gene RFLPs (Pvu II and Hind III) have been associated with hypertriglyceridaemia (Chamberlain *et al.*, 1989), and linked with familial LPL deficiency (Pvu II RFLP; Gotoda *et al.*, 1989). In addition, a gene insertion in the LPL gene has been detected in many families with LPL deficiency (Langlois *et al.*, 1989).

Future prospects

Despite recent efforts, many future studies will be required to characterize genetic factors of atherosclerosis. The baboon model will continue to play an important role in dissection of genetic and dietary effects using studies of baboons of known pedigree fed controlled diets. Studies of human subjects will benefit from linkage analysis to detect cosegregation of gene markers with lipoprotein levels and clinical endpoints of heart disease. New techniques, such as sequencing of PCR-amplified genes from an individual's white cell DNA, will allow characterization of genetic changes that underlie associations between gene markers and disease phenotypes. Ultimately, predictive genetic markers will be developed to determine genotypes for each individual at birth, so that dietary programmes can be tailored to prevent heart disease later in life.

References

Botstein, D., White R. L., Skolnick, M. and Davis, R. W. (1980). Construction of a genetic linkage map in man using restriction fragment length polymorphisms. *American Journal of Human Genetics*, **32**, 314–31.

Breslow, J. L. (1985). Human apolipoprotein molecular biology and genetic variation. *Annual Review of Biochemistry*, **54**, 699–727.

Brown, M. S. and Goldstein, J. L. (1986). A receptor-mediated pathway for cholesterol homeostasis. *Science*, **232**, 34–47.

Bruns, G. A. P., Karathanasis, S. K. and Breslow, J. L. (1984). Human apolipoprotein A-I-C-III gene complex is located on chromosome 11. *Arteriosclerosis*, **4**, 97–102.

Caskey, T. C. (1987). Disease diagnosis by recombinant DNA methods. *Science*, **236**, 1223–9.

Chamberlain, J. C., Thorn, J. A., Oka, K., Galton, D. J. and Stocks, J. (1989). DNA polymorphisms at the lipoprotein lipase gene: associations in normal and hypertriglyceridaemic subjects. *Atherosclerosis*, **79**, 85–91.

Cooper, D. N. and Clayton, J. F. (1988). DNA polymorphism and the study of disease associations. *Human Genetics*, **78**, 299–312.

Davignon, J., Gregg, R. E. and Sing, C. F. (1988). Apolipoprotein E polymorphism and atherosclerosis. *Arteriosclerosis*, **8**, 1–21.

Drayna, D. and Lawn, R. (1987). Multiple RFLPs at the human cholesteryl ester transfer protein (CETP) locus. *Nucleic Acids Research*, **15**, 4698.

Drayna, D., Jarnagin, A. S., McLean, J., Henzel, W., Kohr, W., Fielding, C. and Lawn, R. (1987). Cloning and sequencing of human cholesteryl ester transfer protein cDNA. *Nature*, **327**, 632–4.

Feinberg, A. P. and Vogelstein, B. (1983). A technique for radiolabelling DNA restriction endonuclease fragments to high specific activity. *Analytical Biochemistry*, **132**, 6–13.

Gotoda, T., Senda, M., Murase, T., Yamada, N., Takaku, F. and Furuichi, Y. (1989). Detection of familial LPL deficiency by PvuII RFLP. *Nucleic Acids Research*, **17**, 3607.

Herz, J., Hamann, U., Rogne, S., Myklebost, O., Gausepohl, H. and Stanley, K. K. (1988). Surface location and high affinity for calcium of a 500 kDa liver membrane protein closely related to the LDL-receptor suggest a physiological role as lipoprotein receptor. *EMBO Journal*, **7**, 4119–27.

Hixson, J. E. (1987). DNA markers in primate models for human disease. *Genetica*, **73**, 85–90.

Hixson, J. E. and the Pathobiological Determinants of Atherosclerosis in Youth (PDAY) Research Group (1991). Apolipoprotein E polymorphisms affect atherosclerosis in young males. *Arteriosclerosis and Thrombosis*, **11**, 1237–44.

Hixson, J. E., Borenstein, S. and Cox, L. A. (1990). Pvu II RFLP for the lecithin–cholesterol acyltransferase gene (LCAT) in baboons. *Nucleic Acids Research*, **18**, 384.

Hixson, J. E., Borenstein, L., Cox, L. A., Rainwater, D. L. and VandeBerg, J. L. (1988a). The baboon gene for apolipoprotein A-I: characterization of a cDNA clone and identification of DNA polymorphisms for genetic studies of cholesterol metabolism. *Gene*, **74**, 483–90.

74   J. E. Hixson

Hixson, J. E., Cox, L. A. and Borenstein, S. (1988b). The baboon apolipoprotein E gene: structure, expression, and linkage with the gene for apolipoprotein C-I. *Genomics*, **2**, 315–23.

Hixson, J. E., Kammerer, C. M., Cox. L. A. and Mott, G. E. (1989). Identification of LDL receptor gene marker associated with altered levels of LDL cholesterol and apolipoprotein B in baboons. *Arteriosclerosis*, **9**, 829–35.

Hixson, J. E. and Vernier, D. T. (1990). Restriction isotyping of human apolipoprotein E by gene amplification and cleavage with Hha I. *Journal of Lipid Research*, **3**, 545–8.

Humphries, S. E. (1988). DNA polymorphisms of the apolipoprotein genes – their use in the investigation of the genetic component of hyperlipidaemia and atherosclerosis. *Atherosclerosis*, **72**, 89–108.

Humphries, S., Taylor, R., Jeenah, M., Dunning, A., Horsthemke, B., Seed, J., Schuster, H. and Wolfram, G. (1989). Gene probes in diagnosis of familial hypercholesterolemia. *Arteriosclerosis Supplement I*, **9**, I-59–I-65.

Kowal, R. C., Herz, J., Goldstein, J. L., Esser, V. and Brown, M. S. (1989). Low density lipoprotein receptor-related protein mediates uptake of cholesteryl esters derived from apoprotein E-enriched lipoproteins. *Proceedings of the National Academy of Sciences USA*, **86**, 5810–14.

Langlois, S., Deeb, S., Brunzell, J. D., Kastelein, J. J. and Hayden, M. R. (1989). A major insertion accounts for a significant proportion of mutations underlying human lipoprotein lipase deficiency. *Proceedings of the National Academy of Sciences USA*, **86**, 948–52.

MacCluer, J. W., Kammerer, C. M., Blangero, J., Dyke, B., Mott, G. E., VandeBerg, J. L. and McGill, H. C., Jr (1988). Pedigree analysis of HDL cholesterol concentration in baboons on two diets. *American Journal of Human Genetics*, **43**, 401–13.

MacCluer, J. W., Kammerer, C. M., VandeBerg, J. L., Cheng, M. L., Mott, G. E. and McGill, H. C., Jr (1987). Detecting genetic effects on lipoprotein phenotypes in baboons: a review of methods and preliminary findings. *Genetica*, **73**, 159–68.

McGill, H. C., Jr, McMahan, C. A., Kruski, A. W., Kelley, J. L. and Mott, G. E. (1981a). Responses of serum lipoproteins to dietary cholesterol and type of fat in the baboon. *Arteriosclerosis*, **1**, 337–44.

McGill, H. C., Jr, McMahan, C. A., Kruski, A. W. and Mott, G. E. (1981b). Relationship of lipoprotein cholesterol concentrations to experimental atherosclerosis in baboons. *Arteriosclerosis*, **1**, 3–12.

McLean, J., Fielding, C., Drayna, D., Dieplinger, H., Baer, B., Kohr, W., Henzel, W. and Lawn, R. (1986). Cloning and expression of human lecithin–cholesterol acyltransferase cDNA. *Proceedings of the National Academy of Sciences USA*, **83**, 2335–9.

Mahley, R. W. (1988). Apolipoprotein E: Cholesterol transport protein with expanding role in cell biology. *Science*, **240**, 622–30.

Myklebost, O. and Rogne, S. (1988). A physical map of the apolipoprotein gene cluster on human chromosome 19. *Human Genetics*, **78**, 244–7.

Myklebost, O., Arheden, K., Rogne, S., van Kessel, A. G., Mandahl, N., Herz, J., Stanley, K., Heim, S. and Mitelman, F. (1989). The gene for the human putative apo E receptor is on chromosome 12 in the segment q13–14. *Genomics*, **5**, 65–9.

Paigen, B., Mitchell, D., Reue, K., Morrow, A., Lusis, A. J. and LeBoeuf, R. C. (1987). Ath-1, a gene determining atherosclerosis susceptibility and high density lipoprotein levels in mice. *Proceedings of the National Academy of Sciences USA*, **84**, 3763–7.

Reichl, D. and Miller, N. E. (1989). Pathophysiology of reverse cholesterol transport. Insights from inherited disorders of lipoprotein metabolism. *Arteriosclerosis*, **9**, 785–97.

Rigby, P. W. J., Dieckmann, M., Rhodes, C. and Berg, P. (1977). Labeling deoxyribonucleic acid to high specific activity in vitro by nick translation with DNA polymerase I. *Journal of Molecular Biology*, **113**, 237–51.

Saiki, R. K., Scharf, S., Faloona, F., Mullis, K. B., Horn, G. T., Erlich, H. A. and Arnheim, N. (1985). Enzymatic amplification of beta-globin genomic sequences and restriction site analysis for diagnosis of sickle cell anemia. *Science*, **230**, 1350–4.

Small, D. M. (1987). HDL system: a short review of structure and metabolism. *Atherosclerosis Review*, **16**, 1–8.

Southern, E. M. (1975). Detection of specific sequences among DNA fregments separated by gel electrophoresis. *Journal of Molecular Biology*, **98**, 503–17.

Sudhof, T. C., Goldstein, J. L., Brown, M. S. and Russell, D. W. (1985). The LDL receptor gene: a mosaic of exons shared with different proteins. *Science*, **228**, 815–22.

Tall, A. R. (1986). Plasma lipid transfer proteins. *Journal of Lipid Research*, **27**, 361–7.

Wion, K. L., Kirchgessner, T. G., Lusis, A. J., Schotz, M. C. and Lawn, R. M. (1987). Human lipoprotein lipase complementary DNA sequence. *Science*, **235**, 1638–41.

4 *Hypervariable minisatellites and VNTRs*

MARK L. WEISS AND TRUDY R. TURNER

Introduction

The human genome contains several families of highly repetitive DNA sequences, known as satellite DNA. These repeated elements can be divided into a variety of classes based on the number and organization of the repeated unit. One such class was described by Jeffreys and colleagues (1985a) as 'minisatellite' DNA; while satellite DNA consists of many copies of a short sequence interspersed in the human genome, the minisatellites consist of a smaller number of tandem repeats.

Minisatellites found in the human genome exhibit exceptionally high levels of variability and highly polymorphic minisatellites have been detected in a number of non-human primate species. Although there are only hints as to the possible functions for the minisatellites, their hyper-variability provides a very useful tool for investigating a wide range of practical and theoretical questions, e.g. in phylogeny, in primate social structure, and in human identification and other forensic settings (Gill *et al.*, 1985; Jeffreys *et al.*, 1985c; White and Greenwood, 1988; Acton and Harman, 1989).

Classical genetic markers such as blood groups, serum protein and red cell enzymes were formerly utilized, for example, in estimating genetic variability, tracing genetic relationships of conspecific or congeneric populations, demonstration of the operation of evolutionary forces, human identification through the use of genetic markers, uncovering linkage relations between markers and genetic disorders, resolution of questioned parentage, and management of captive colonies and populations. For all these, more refined analysis is now possible using the tools of molecular biology and the greater variation that they reveal. Restriction fragment length analysis and the polymorphisms so identified (see Chapter 2) have already proved to be highly useful in linkage group determinations (Barker *et al.*, 1987; Lander and Botstein, 1987; White *et al.*, 1989), prenatal disease diagnosis (Boehm *et al.*, 1983), and population studies (Ewens *et al.*, 1981; Wainscoat *et al.*, 1986; Cann *et al.*, 1987). However, such RFLPs have one disadvantage. As they can only be

76

present or absent, there are only two alleles and hence the maximal level of heterozygosity is limited (by the allele frequencies) to 50%. To a degree, this may be modified if several linked, polymorphic restriction sites are tested simultaneously, but this too can be cumbersome as it requires multiple digests.

Hypervariable loci

Initially, RFLPs do not appear to provide a class of variants which demonstrates 'hypervariability'. Such loci would contain many alleles, each present at low frequency in the population. In 1980, Wyman and White detected a highly polymorphic locus. At the time, their probe was anonymous in that it recognized a DNA sequence without a known association or function; it is now known to be derived from a region of chromosome 14 close to the immunoglobulin genes (Balazs *et al.*, 1982). When hybridized to human DNA cut with Eco RI at least eight alleles were discovered in the 43 individuals tested. When cut with another restriction enzyme, a second polymorphic locus with two alleles was revealed, for a total of 16 allelic combinations. As Wyman and White noted, the structural basis for the polymorphism is the number of repeats of a short sequence of DNA. This is a second class of RFLPs which provide evidence for a large number of alleles and the absence of one ubiquitous allele.

Series of loci sharing these features were characterized in the human genome by A. J. Jeffreys and colleagues (Jeffreys *et al.*, 1985a, 1985b, 1990; Wong *et al.*, 1986, 1987; Royle *et al.*, 1988; Armour *et al.*, 1989b). The probes used to detect the loci are based upon a 33 bp sequence originally encountered in the second intron of the human myoglobin gene. This short sequence was known to be repeated in tandem (Weller *et al.*, 1984) to produce a 'minisatellite'. Genomic DNA, digested with a restriction enzyme which did not cut within the 33 bp repeat unit, was probed. The resulting pattern gave evidence for a number of fragments, each corresponding to a genomic segment containing a different number of tandem repeats of a sequence related to the 33 bp unit. The bands appeared to be inherited in a Mendelian fashion.

Using the original 33 bp sequence as a probe, a human genomic library was screened and a number of randomly selected cross-hybridizing recombinant phage were isolated under low stringency conditions; that is, conditions which allow hybridization of probe and target to take place in spite of some degree of base mismatch. Sequence analysis showed that each contained a number of repeats of a sequence related to the 33 bp probe. Embedded in the repeat unit is a core, or conserved, consensus sequence common to all the cross-hybridizing sequences. In some, the repeated sequence is 16 bases long and corresponds closely to a conserved sequence of c. 15 bp found in the myoglobin repeats. In others, the repeat is roughly twice that length and in yet others, four times. In all cases, the repeats contain a version of this 15–16 bp conserved core plus other sequence. In addition, the tandem repeats are bounded by highly divergent flanking sequences.

Probes were prepared from each of these cloned tandem repeats and hybridized to digested human DNA under conditions requiring a high degree of probe and target sequence complementarity. Each of eight probes identified a unique region of the human genome and four of the eight probes detected polymorphic variation in a panel of unrelated individuals. Pedigree analysis indicated that the hybridizing bands are inherited in a Mendelian fashion and that they exhibit somatic and germ-line stability (Jeffreys *et al.*, 1985a). Thus, there is a basis for uncovering large amounts of variability rapidly, hence the phrase 'hypervariable minisatellites'. Focusing on their structural characteristics, Nakamura *et al.* (1987) used the term 'variable number of tandem repeats' or VNTRs for these probes.

The multiple bands revealed by minisatellites represent alleles at a number of independent loci. Some alleles are lost in the large number of small fragments seen at the bottom of the autoradiographs, but Jeffreys *et al.* (1986) estimated that using the two probes known as λ-33.6 and λ-33.15 approximately 60 hypervariable loci can be visualized simultaneously in human DNA. In any given pattern, however, only half this number of loci will be scorable. Thus, in comparing the patterns of two individuals one is usually looking at overlapping but different arrays of loci. This negates the ability to conduct standard population genetic analyses as performed with classic genetic markers.

Two features are observed when comparing such patterns: first, the electrophoretic mobility and secondly, the intensity of the band. The ability to differentiate mobilities is, of course, in part a function of visual acuity, and constancy of electrophoretic phenomena. Generally, one can detect differences in mobility in the range of 0.5 mm. For a gel separation of 15 cm from the largest to the smallest scorable fragments, there are c. 300 resolvable band states. While legal consequences demand absolute certainty in discriminating between band states (Reilly, 1990), establishing a reasonable equivalence of bands for population and primate parentage analysis is rarely a problem.

Because of the large number of loci which Jeffreys's initial probes concurrently screen and the hypervariable nature of at least some of these loci, the likelihood that two individuals will show the same banding pattern is exceptionally low (Jeffreys *et al.*, 1985b). The likelihood that sibs in a non-consanguineous marriage will share a particular fragment is 50%. Thus, with 15 scorable bands,the probability that sibs will have identical patterns is 2^{-15}. Using both probes, the average number of scorable bands per individual is about 36, leading to a probability of identical patterns in sibs of 3×10^{-14} (Jeffreys and Morton, 1987). For unrelated individuals in British and Indian populations, the mean level of band sharing is about 25%. Thus, with 36 resolvable fragments the probability that all bands seen in one individual will be seen in a second, unrelated person is 0.25^{-36} or 2×10^{-22} (Jeffreys, 1987). Given probabilities in this range, the patterns can reasonably be considered individual-specific and thus have come to be called DNA 'fingerprints' or 'profiles'. Many of these points are expanded upon in Kirby (1990). Lately, some have extended this term to unique banding patterns detected with a diversity of probes in either nuclear or mitochondrial DNA (Avise *et al.*, 1989).

Single locus probes

The ability to screen a large number of loci with one sweep has drawbacks in certain applications. It produces a unique pattern of bands, but it is difficult to determine which bands are alleles. In situations of questioned parentage or other forensic applications this is of little consequence. For population analysis, it is of major significance. If necessary, individual bands can be isolated from a fingerprint and used as a probe. The utility of this approach has been demonstrated by Wong *et al.* (1986), who purified and sub-cloned a fragment that reacted with probe λ-33.15 and appeared to segregate with HPFH (hereditary persistence of fetal haemoglobin) in a large pedigree. The resulting clone had about 170 tandem repeats of a 37 bp core sequence. The repeats were not totally homogeneous; some had a 4 bp deletion and others had a variety of base-pair substitutions relative to the consensus sequence. Internal to the repeat unit was a 12-base sequence bearing strong homology to the core sequence found in all the minisatellites in this series. Thus, when the cloned fragment was hybridized to genomic DNA at high stringency, it reacted with one or two fragments in all individuals, the former presumably represent homozygotes, the latter heterozygotes. Probes which detect only one or two bands per individual are 'single-locus' probes in contrast to multi-locus or polycore probes. The inability of such probes to hybridize to other fragments results from the presence of non-core DNA and disturbed or offset phasing of the repeat units. Fragments detected with the single-locus probes are demonstrably Mendelian in character and exhibit germ-line stability.

A large battery of probes is now available to detect genetic markers. In addition to Jeffreys's probes, other have been developed, e.g. minisatellite tandem repeat sequences from insulin (Rotwein *et al.*, 1986), apolipoprotein B (Knott *et al.*, 1986; Ludwig *et al.*, 1989) apolipoprotein CII (Rogaev, 1989) and collagen type II (Stoker *et al.*, 1985). A number of polymorphic minisatellites have been uncovered with probes from the α-globin gene complex, including sequences from the vicinity of the α1 gene (Jarman *et al.*, 1986), the 5' end of the complex (Jarman and Higgs, 1988), the 3' end of the cluster (Fowler *et al.*, 1988) and ζ intron 1 (Goodbourn *et al.*, 1983). Devor and Burgess (1989) constructed two oligonucleotide probes based on four base repeats initially discovered in other vertebrates. Tandem repeats from the virus M13 (Vassart *et al.*, 1987) also serve as probes of the human genome. Vergnaud (1989) used random sequence 14-mer oligonucleotide probes to detect polymorphic loci. Nakamura *et al.* (1988) also synthesized probes, but they specifically included the sequence GNNGTGGG, a consensus sequence found in a number of the probes used in their earlier work.

In fact most, but not all, of the minisatellite core sequences appear to be GC-rich and may form families of related sequences (Jarman and Higgs, 1988). Some of the informative probes resemble another GC-rich sequence, Chi, which is a signal for recombination in bacteria. This suggests that unequal crossover in recombination 'hotspots' may be a process by which the number of repeats may increase or decrease. The suggestion is supported by the discovery that minisatellites cluster in the preterminal regions of human autosomes (Royle *et al.*, 1988), the same regions that show preferential clustering of recombination nodules in male meiosis.

Screening 78 randomly chosen individuals from the British population with Wong's original single-locus probe revealed at least 77 alleles each having between 14 and more than 500 repeats. The most frequent was present at 16% frequency, the other 76 were quite rare. The expected heterozygosity was greater than 96% for this one locus and in fact only four homozygotes were found. The same approach produced a number of additional locus-specific probes based on fragments originally detected with probes λ-33.15 and λ-33.6. For each, the heterozygosity ranges from 90 to 99%.

In a novel approach to genotyping, Jeffreys *et al.* (1990) use a system based on partial double digests to generate as many as 10^{70} potential allelic states, in contrast to the 300 mentioned earlier. Here, the pattern of variable restriction sites within repeats is mapped so that bands of similar size can be differentiated. Given such discriminatory power, it is not surprising that the heterozygosity levels appear to approach 100%.

An alternative approach to increasing the discriminatory power was described by Uitterlinden *et al.* (1989). After separating the digested DNA by size in a neutral 6% polyacrylamide gel, a size-selected region of the gel was excised and analysed by electrophoresis through a second dimension in a denaturing polyacrylamide gel. The second step serves to separate fragments according to their AT:GC ratio. Two-dimensional representations, while generating up to 625 spots, can be rather difficult to score between gels.

Other probes reveal high levels of variability in humans as well. The ten oligonucleotide probes which Nakamura *et al.* (1987) employed revealed 77 new VNTR loci. Of these, 67 had three or more alleles and averaged more than 70% heterozygosity. Tynan *et al.* (1990) found 32 fragments among 122 people of widely scattered backgrounds. Balazs *et al.* (1989) screened a much larger series of human samples from several ethnic groups with five probes, some of which were also utilized by Nakamura *et al.* (1987). While each locus had at least 30 alleles there were significant differences between groups in the distribution of alleles and there were differences in levels of heterozygosity between loci. A recent compendium provides many useful population data for forensic scientists (International Symposium on Human Identification, 1990).

The generation of new alleles

'Slipped strand mispairing' (SSM: Levinson and Gutman, 1987) is one mechanism by which new size variants can be generated. Local denaturation of the strands of DNA duplex followed by mispairing of complementary bases at the site of a tandem repeat can lead to insertion of a copy of the repeat unit upon replication. A comparable situation can lead to the loss of a repeat unit. Other mechanisms are possible whereby the number of units can be increased or decreased. In humans, Jeffreys *et al.* (1988a) found that the mutation rate at the locus detected by λ-MS1 was equal in sperm and oocytes. This would imply that the mutation rate is not cell-division dependent and would argue against SSM as the predominant mechanism for generation of new size variants. However, the size distribution of mutations differs significantly in sperm and oocytes; sperm show an excess of small mutation events. The small size and high mutation rate in sperm (which undergo many more cell divisions than oocytes) are consistent with SSM. Large mutations generally involving gain or loss of up to 200 repeat units appear to be replication independent and may arise through unequal crossovers or gene conversion at meiosis. This is consistent with the idea previously mentioned that the core sequence is a human recombination signal. Armour *et al.* (1989a) found that large changes can occur in tumour cells in the absence of meiosis. More recently, Jeffreys *et al.* (1990) investigated one minisatellite and found no evidence for interallelic unequal exchange as a mechanism for generating size variants. In fact, their most recent data argue strongly against this minisatellite serving as a recombination hotspot involved in chromosome pairing or meiotic recombination. The apparent contradiction may relate to phenomena which differentially affect repeat sequences of different sizes (Stephan, 1989) or different base composition.

At another level the 'cause' of minisatellites and their wealth of alleles focuses on function. Jeffreys *et al.* (1985a) initially suggested that minisatellites served as hotspots for recombination events and there are some data on recombination levels in the vicinity of minisatellites to support the view (Jarman and Wells, 1989). At the moment, it is simply unknown whether the minisatellites serve this or any other function. However, the fact that Collick and Jeffreys (1990) found a ubiquitous minisatellite-specific DNA binding protein suggests that minisatellites serve as a recognition signal.

Determination of parentage

The germ-line stability and the Mendelian inheritance of minisatellite alleles allows for the identification of parentage in humans (Jeffreys *et al.*, 1986). Barring mutations, all bands present in a child can be traced to either the mother or the father. In the case of questioned parentage, most usually of the father, fragments common to the offspring and the known parent can be disregarded and the offspring and suspected parent compared for all other bands. If bands are present in the child which are not found in the suspected parent, then the case against the suspect can be dismissed. Given the low frequencies of specific fragments, statistical proof of parentage may be accomplished if all informative bands in the child can be traced to the suspect. The statistical power of the test is so great that it is possible to perform parentage inclusions, as opposed to the classic exclusions. One can, with statistical certainty, determine if a child is the offspring of one individual or of that individual's sibling (Jeffreys *et al.*, 1985c).

In spite of the tremendous power of the techniques, much has been made in the media of the 'failings' of DNA 'fingerprinting' in forensic settings. There are certainly points to consider before rushing to unquestionable acceptance of the results (Marx, 1988; Lander, 1989; Lewin, 1989; Reilly, 1990). In fact, the science of the technique is not the problem. The proper handling and administration of the samples, the use of appropriate controls, the skill of the laboratory personnel and the ethical management of the results must be properly monitored.

Applications in non-humans

At least some of these probes can be hybridized to the DNA of non-human primates as Weiss *et al.* (1988) and others (Dixson *et al.*, 1988; Tynan and Hoar, 1989; Washio *et al.*, 1989; Ely and Ferrell, 1990) have demonstrated for several primate species. As such the technique has a number of potentially important practical applications. Additionally, DNA profiles can help to clarify thus far elusive evolutionary relationships between genes and behaviour.

Captive populations

Captive colonies of primates are found primarily in zoos and in biomedical or behavioural research facilities. In recent years zoos have undergone a partial transformation, from facilities where animals are kept primarily for display and education to research-orientated refuges for endangered populations. Zoo personnel have for years recognized the advisability of outbreeding to maintain the viability of their populations. However, it has been in the last several years that the ISIS system, an international breeding record for selected species, has made some controlled breeding possible. Zoos often use conspecifics of unknown and possibly widely divergent provenance to populate a colony. In the past, zoos selected members of a colony without regard to genetic constitution: after a colony was established there was usually little immigration. Wild populations of primates often have low levels of heterozygosity (Melnick and Pearl, 1987); however, widely separated local populations can have highly divergent allele frequencies (Fooden and Lanyon, 1989). Even if the zoo's founding population is constituted so as to maximize heterozygosity, the small size of most zoo colonies and the lack of immigration can lead to high levels of inbreeding and soon greatly reduce the initial genetic variability.

To what degree is it desirable to maximize genetic variation? Natural populations have evolved social and genetic mechanisms that balance inbreeding and outbreeding. There have been several studies attempting to clarify the amount of inbreeding and outbreeding in natural populations. Studies of bears (Rogers, 1987), wolves (Mech, 1987) and prairie dogs (Halpin, 1987) among others indicate that these populations are actually quite inbred. Shields (1987) suggested that while there may be some costs of extreme inbreeding, these costs are often balanced by the social benefits of philopatry. The need to maintain levels of heterozygosity in populations must also be balanced by the need to avoid outbreeding depression, which can result if mating occurs between members of groups that would never mate in the wild.

Biomedical research facilities also maintain colonies of primates. Until the last decade most of these animals were obtained from commercial sources. However, since the 1970s, countries with indigenous primate populations have curtailed animal exportation. Biomedical research facilities must now breed animals and maintain viable breeding colonies. These facilities need information about genetic and epidemiological factors that may affect the health of the colony. In some cases, animals are bred for their susceptibility or resistance to disease. Paternity assessment and the presence of genetic markers are essential for this type of investigation. Depending on the way in which primate populations are housed, behavioural data may be collected as well. If populations are allowed to form normal social groups and have significant space to range, they can be monitored for social interactions. Information on captive colonies housed in this manner can contribute to the answering of theoretical questions about kin selection. However, this also necessitates knowing the parentage of the animals.

Unlike the situation in the wild, the number of potential fathers in a captive environment can be limited and controlled. Smith (1981) and Duvall et al. (1976) attempted to ascertain paternity in non-human primates on the basis of an electrophoretic examination of blood proteins. In both cases, fathers were selected from a limited pool of males and mothers were known. Smith (1981) identified 202 of 220 fathers of rhesus (*Macaca mulatta*) infants born at the California Regional Primate Center and Duvall et al. (1976) identified 25 out of 29 infants at Yerkes Regional Primate Center. Paternity determination using electrophoresis requires a large number of exclusion tests since phenotypes have to be established independently for each blood protein system. Smith (1981) performed 478 of the 492 exclusions needed for full identification, while 203 exclusions would have been needed by Duvall *et al.* (1976) for full exclusion.

Jeffreys (1987) initially determined that probes λ-33.6 and λ-33.15 cross-hybridized with a variety of vertebrates including mammals, birds, amphibians and fish. The utility of DNA analyses in animal husbandry was readily apparent, for animal identification, verification of semen samples for artificial insemination, and linkage analysis of economically important traits. Jeffreys and Morton (1987) examined eight breeds of dogs and five breeds of short-haired domestic cats and found substantial variability, even though there was greater band sharing than is found in humans. They found the patterns of some farm animals, including sheep, goats, pigs and cows, less informative. Georges *et al.* (1988a), however, found extensive variability in cattle using additional probes including M13mp9, the 3'α-globin HVR probe and the Per probe. They estimated heterozygosity of 0.89–0.92 in 16 animals of known relationship. Known families of five horses (Belgian halfbreed), four dogs (beagles), four crossbred pigs, three chickens and a group of barbels (a European freshwater fish) were examined and found to have individual-specific DNA patterns with the M13, Jeffreys and Per probes. Heterozygosities were calculated as 0.66–0.85 in horses, 0.85–0.86 in dogs and 0.76–0.79 in pigs. These values are lower than those of humans but Georges *et al.* (1988a) calculated that they are high enough to determine paternity if the real and the putative father are not related. An example of this is a case of disputed canine paternity (Georges *et al.*, 1988b). A Shui Tzu bitch was accidentally inseminated by two dogs, one a Coton de Tulear. DNA profiles were used to determine that one of the two pups produced was a purebred and the other a mixed breed.

DNA patterns are being used in the genetic management of a number of captive species. The San Diego Zoo has analysed all 28 California condors held for breeding in captivity. Future matings of this highly endangered species will use DNA analysis as a tool to avoid inbreeding. It is also being used on the zoo's Galapagos tortoises. The National Zoo in Washington, DC, is using fingerprinting on two endangered species of birds from Guam, the rail and the kingfisher. There has been considerable success in the breeding of rails and eventually they will be reintroduced into the wild. The St Louis Zoo is attempting to avoid outbreeding depression in African elephants and other species where wild local populations are often widely separated (Weiss, 1989). Turner and Weiss have used probes λ-33.6 and λ-33.15 to screen a number of *Papio papio* from the Brookfield Zoo baboon colony. This population has been in existence for a number of decades and until recently received little genetic management. Figure 4.1 shows the effects of generations of inbreeding on four members of the colony chosen at random. While the complexity of the pattern is reasonably high, the distinctive nature of fingerprints has largely vanished, attesting to the increased homozygosity associated with inbreeding. DNA fingerprints of feral baboons (*P. cynocephalus*) show a much more varied gene pool. The diminished variability of the zoo population is neither characteristic of baboons, nor an artefact of hybridization.

Three colonies of *Macaca sylvanus* (Barbary macaques) are maintained in France and Germany. Figure 4.2 shows the DNA fingerprints for a number of the animals in the Rocamadour enclosure which now numbers over 200 animals. These 'unrelated' animals share about 45% of their bands. This is roughly the same level of band sharing as is seen in a small sample of dogs of different breeds and noticeably greater than the 25% seen in humans (Jeffreys *et al.*, 1986). Since pedigrees on this Barbary macaque population were not maintained and population data from feral groups do not exist, it is impossible to tell whether the high level of band sharing is characteristic of the species or a result of the population's history. One can say that on the original autoradiograph (probed with λ-33.6) the average animal had eight scorable bands. This seems to be a species characteristic and, from experience with the system in the genus *Macaca*, appears to be about average. Assuming that all cases of band sharing result from identical alleles at the same locus, then the mean allele frequency of minisatellite fragments is about 26%. If it is assumed that (1) comigrating bands are identical alleles and (2) there is a low variance in allele frequencies, then this is also the mean homozygosity. This is an overestimate for those cases where comigration of different bands arises fortuitously. From the Kintzheim macaque colony the probability that two randomly selected unrelated individuals are identical in all bands is about 4×10^{-6}. While this is a far higher probability than seen in humans, it is still quite rare and is thus of potential use to those responsible for primate colony management.

Wild and semi-wild populations

Early studies of the genetics of primate populations concentrated on systematics and interspecific and interpopulation differences in gene frequencies. Several studies of wild populations conducted in the 1970s and the 1980s attempted to determine which stochastic or deterministic models of evolutionary processes were instrumental in producing the

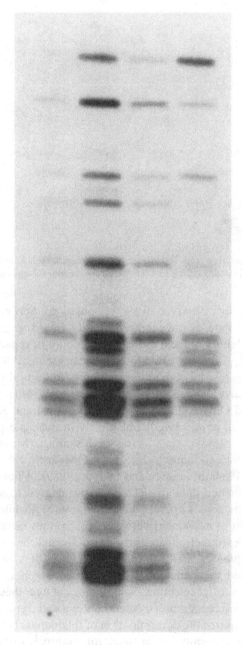

Fig. 4.1. DNA fingerprints of 4 randomly chosen baboons from the Brookfield Zoo colony. In all figures, the genomic DNA was cut with Alu I and hybridized to either λ-33.6 or λ-33.15.

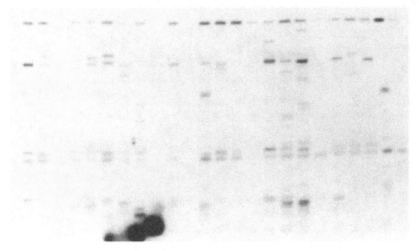

Fig. 4.2. DNA fingerprints from 25 *Macaca sylvanus* from La Montagne des Singes. Weakly hybridizing patterns are visible on the original autoradiograph.

pattern of genetic variability observed in local populations. Serogenetic data were used to calculate average heterozygosity (H), the genetic distances and genetic identity of local groups, the amount of separation between populations (F_{ST}) and migration rates between groups. All these studies were on semi-terrestrial cercopithecoids, e.g. in natural populations of baboons in Kenya and Ethiopia (Olivier *et al.*, 1974; Ober *et al.*, 1978; Coppenhaver and Olivier, 1986), vervets in Ethiopia and Kenya (Turner, 1981; Dracopoli *et al.*, 1983), but mostly in macaque populations and especially the rhesus population on Cayo Santiago (McMillian and Duggleby, 1981; Buettner-Janusch *et al.*, 1983; Ober *et al.*, 1984) and in Asia (Melnick *et al.*, 1984a,b; Fooden and Lanyon, 1989). All of these studies used gene frequency data to determine the degree to which stochastic or deterministic forces operate on populations but, while they can estimate the amount of movement between groups, they cannot be used to determine the reproductive success of any individual.

E. O. Wilson's *Sociobiology* (1975) shifted the emphasis of evolutionary studies and included the behavior as a quantifiable genetic trait subject to the action of natural selection. Sociobiological hypotheses by definition require some kin recognition by members of a social group. To test these hypotheses requires the determination of relatedness between individuals, particularly paternity, to answer, for example, questions such as: What is the reproductive success of high ranking males? Does reproductive behaviour match reproductive success? What is the actual genetic contribution of a male during his tenure in a one-male group? How much inbreeding actually occurs? In cases of infanticide, do males preferentially kill the offspring of non-relatives? Do they then sire the next infant? Behavioural observation can provide information on the mother of an infant, on priority of access of resources, on the amount of physical contact and social relations between individuals. Moreover, electrophoretic analysis of blood proteins is rarely sufficiently comprehensive to do so. All primate groups examined have low levels of heterozygosity. Average heterozygosities (H) for 15 macaque species range from 0.018 to 0.1772. The highest levels are found in the Indonesian species, *Macaca tonkeana*, *M. maura*, *M. hecki*, *M. nigra* and *M. nigrescens*. Ciani *et al.* (1989) suggest the presence of hybrid zones and gene flow between these populations which would explain the high levels of variability. The three species that have been examined most closely, *M. fuscata*, *M. mulatta* and *M. fasicularis*, all have average heterozygosities that range from 0.012 to 0.108. Populations of *Papio anubis*, *P. hamadryas* and *P. cynocephalus* have average heterozygosities of 0.019–0.096. Vervet monkeys (*Cercopithecus aethiops*) are the least variable, with levels of 0.021–0.05. Macaques seem to be the most diverse in species number, adaptation and gene frequency levels. They are followed by baboons and then vervets (T. Turner, unpublished data). By contrast DNA analysis, because of the high levels of detectable heterozygosity, will help answer such questions. Analysis in combination of restriction fragment length polymorphisms, hypervariable minisatellites and VNTRs would be the best strategy for screening populations, especially when samples are small and reacquisition is impossible. DNA 'fingerprinting' has been used to obtain estimates of heterozygosity in several natural populations. The largest population sampled is human. Jeffreys (1985a) examined a Gujerati Indian kinship with 54 members spanning four generations. Forty large three-generation families comprise the CEPH (Centre d'Etudes de Polymorphisme Humain) reference panel (Jeffreys *et al.*, 1988a; Balazs *et al.*, 1989). Additional large multi-generation families comprise the Mormon kindreds of Utah (Nakamura *et al.*, 1988). Heterozygosity levels were found to approach 100%.

A wild population of house sparrows (*Passer domesticus*) near Nottingham, UK under observation since 1979 shows extremely variable band patterns. Mother/son incest was detected. In addition, not all members of a single brood had the same father. Approximately 8% of the nestlings are genetically mismatched with one putative parent (Wetton *et al.*, 1987). Five species of birds were examined by Burke and Bruford (1987). In the sparrow population they examined, it was again possible to detect different fathers for different members of a brood. Additionally there was evidence of intraspecific brood parasitism where a female laid her eggs in the nest of another female. Burke *et al.* (1989) reported on a population of dunnocks (*Prunella modularis*). About 80 birds in the Cambridge University Botanic Garden evidenced a variable social organ- ization including monogamy, polyandry, and polygyny. DNA profiles showed that all but one of the 133 offspring were fathered by a resident male and there were no cases of intraspecific brood parasitism. Male dunnocks helped feed the chicks and polyandrous males were more likely to feed chicks if paternity were possible. It was clear, however, that the males did not recognize their own offspring since they fed chicks they had not fathered. Males seemed to use mating access to determine if they would feed the young. A male's greatest reproductive success is in a monogamous relationship. However, chicks had the best chance of survival in a polyandrous situation. The use of DNA analyses helped in the explanation of this complicated social structure.

Skin biopsies of members of a population of killer whales (*Orcinus orca*) were obtained and DNA analysed. Hoelzel and Amos (1988) argued that to ensure conservation of this species adequate census data are needed. This information can be provided by DNA multilocus probe analysis as shown in five other cetacean species.

While not a wild population, the semi-wild population of *Lemur catta* at the Duke Primate Center has been analysed for parentage by Pereira and Weiss (1991). This was a particularly interesting case as it was possible to investigate the role of female choice in avoiding inbreeding. To demonstrate inbreeding avoidance, it is necessary to know both the degree of genetic relatedness in all possible male–female pairs and the identities of all pairs that reproduce. Previous research in this area often lacked precise information on these variables. In contrast to other primates, female mating preferences in ringtailed lemurs are relatively easy to observe because all adult females agonistically dominate all adult males and because female sexual receptivity is restricted to oestrous periods lasting for 8–12 hours. DNA fingerprinting analyses allowed the determination of the 17 reproductive pairings that occurred in this colony over a 5 year period.

Blood samples were gathered from all animals represented in this study except two. DNA from one was prepared from a frozen kidney; DNA was not available from the other, a male and a potential father to some of these animals. The hope was that any offspring sired by this male would be identifiable by exclusion of all other potential fathers. High molecular weight DNA was prepared and digested with Alu I using standard techniques. It was analysed by electrophoresis and transferred to nitro-cellulose filters, then probed with λ-33.6 and λ-33.15.

All lemur samples yielded informative DNA fingerprints. A female, her offspring and four potential fathers are shown in Fig. 4.3. The patterns exclude all but male 4 from possible paternity. For 15 of the 17 infants under study, all but one male could be excluded from paternity. In the remaining two cases, all males for which DNA was available were excluded and paternity was assigned to a single deceased male. This paternity assignment was internally consistent as the two offspring shared paternally-derived bands. A control case for paternity assignment was provided by two matings for which colony management records allowed only one possible sire. In both, the DNA fingerprinting was in agreement with records, as shown in Fig. 4.4.

With paternity information to hand, it was possible to document inbreeding avoidance by the females. Pooling parentage information, it was demonstrated that from 1982 to 1986 inclusive, four of the five reproductive females always mated with males likely to share 25% or fewer of their genes or with the otherwise least-related available male. There were 15 conceptions which occurred in circumstances that allowed the female to choose between closely and distantly related males; in 12 the female chose a male who shared 25% or fewer of her genes. No female ever reproduced with a son or matrilineal brother. Only one female ever bred with a matrilineal nephew. Two of three females having opportunities to breed with their fathers avoided ever doing so; however, the other mated with her father twice in five opportunities. The four other relatively inbred matings involved a matrilineal cousin. In two cases, this male was the older of the two least-related males available. He was also the only patrilineal half-brother to breed ($n = 4$). In sum, the females clearly avoided breeding with close matrilineal kin, but insufficient evidence was obtained to show avoidance of mating with patrilineal kin.

On two occasions, unrelated males were introduced to simulate male immigration and diversify female mating opportunities. While the resident males attacked the potential immigrants for several months at least five adult females engaged the males in mutual grooming during observations. With DNA fingerprints it was shown that immigrant males who had remained bottom-ranking in the male hierarchy did nevertheless impregnate a disproportionately large number of females.

Thus, with the information provided by DNA fingerprinting, one can conclude that female mating preferences and avoidance of inbreeding can override the dominance relations among males to influence male mating success.

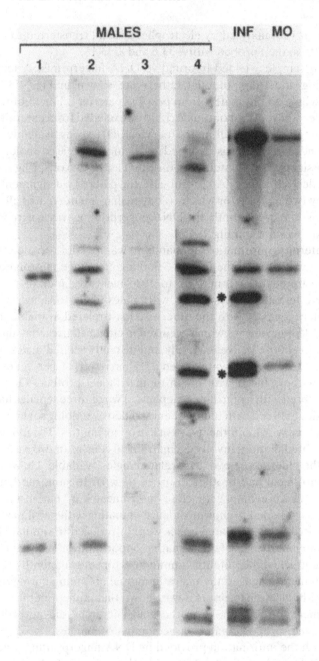

Fig. 4.3. Ringtailed lemur DNA fingerprints exclude all pictured males except animal 4 from paternity of the infant. The informative bands are indicated.

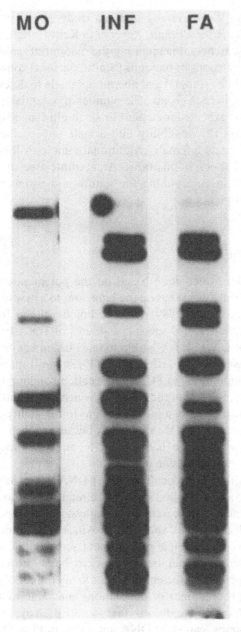

Fig. 4.4. Colony management records indicate that the male (FA) must have been the father of the infant ringtailed lemur. DNA fingerprinting does not exclude this male from paternity.

Currently, Turner and Weiss are engaged in a study of wild primate populations. In the Tana River Primate Reserve in Kenya, there are at present behavioural researchers documenting the amount of male migration between four local troops of baboons (Smith, personal communication). These troops will be trapped and an attempt made to determine the paternity of all young in each group. The number of potential fathers is not as great as might be expected for a multi-male, multi-female troop. However, there is always the possibility that a male outside the area may move in and impregnate a female. Additional samples will be collected from the Awash baboon populations. An accurate assessment of paternity here will help in determining the nature and extent of the hybridization in these populations.

Directions for the future

In the past several years a technique based on the polymerase chain reaction or PCR (Saiki *et al.*, 1985) has opened the door to a new wave of investigations (cf. Pääbo *et al.*, 1989). This technique allows for the amplification of large quantities of specific segments of DNA from as little as one molecule of DNA. It is possible to use PCR in conjunction with the analysis of minisatellites as Jeffreys *et al.* (1988b) and others have done. There are some limitations to PCR amplification common to any PCR usage, for example, the size of the fragment which can be faithfully amplified. Additionally, there was some concern as to the ability of the technique to amplify minisatellite DNA faithfully because of the tandem repeat structure. Neither concern has proved to be a problem and minisatellites of up to 10 kb have been amplified.

The ability to start with one molecule of target DNA and proceed to grow sufficient copies to allow detailed analysis is already revolutionizing many areas within biomedical research. Within physical anthropology, the coupling of single-locus probes and PCR provides opportunities as numerous as they are exciting. Sample procurement and handling will be greatly eased. For example, a number of plucked hairs, stored and shipped in ethanol can provide sufficient DNA for many rounds of PCR, thus greatly expanding the number of populations amenable to study. Blood samples, stored under conditions that would destroy red cell enzymes and other proteins, can yield DNA amenable to PCR (Towne and Devor, 1990). Likewise, as DNA is a rather stable molecule, samples from long deceased individuals can provide historical genetic data; even formalin-fixed, paraffin-embedded samples can be used (Pääbo *et al.*,

1989). The understanding of human population genetics, genetic adaptation, and the structure of human populations (Boerwinkle *et al.*, 1989; Vigilant *et al.*, 1989) is already beginning to reap significant benefit from the application of PCR to the study of minisatellite loci.

For primatologists there are many opportunities too. The most exciting may be the complete ascertainment now readily possible without use of invasive techniques. Again, the use of hair bulb DNA can provide sufficient material to characterize primate populations. Specimens stored in a dry state (e.g. museum specimens) also can yield usable DNA (Kocher *et al.*, 1989).

There are still several factors restricting the application of the technique to non-human primates. The rapid rate of minisatellite evolution (Jeffreys *et al.*, 1988a, 1990; Nurnberg *et al.*, 1989) prevents the use of human single-locus probes on all but our closest relatives; some single-locus probes cross-hybridize with chimpanzee and gorilla DNA (M. L. Weiss and T. Turner, unpublished data), but that is the extent of their utility. As single-locus probes are developed for non-human primate species, the ability to gather genetic data on primate colonies and populations will increase dramatically.

Some have attempted to determine degrees of relatedness between animals in natural populations, but this extends the data beyond their limits (Lewin, 1989a,b). Yet, there is no doubt that the information being gathered will greatly enhance our knowledge of the relationship between primate behaviour and evolution.

Applied with an appreciation for its potentials and limitations, DNA fingerprinting can help to answer questions across many areas of physical anthropology and primatology. The future will see the application of DNA fingerprinting techniques to the clarification of a broad range of theoretical and practical questions, both within molecular anthropology and in the broader spectrum of behavioural sciences, as a means of elucidating the interplay of behaviour, genetics and evolution.

Acknowledgments

The authors thank A. J. Jeffreys for use of the polycore probes. He and his associates have provided much assistance and support over the years. The minisatellite probes are subjects of Patent Applications: address commercial inquires to ICI Diagnostics, Gadbrook Park, Rudheath, Northwich, UK. We are deeply indebted to Michael Pereira (DUPC) for information on the behaviour and history of the Duke ringtailed lemur colony; many of the data discussed here are drawn from Pereira and Weiss (1991). We also thank the following for providing samples: Robert Lacy (Brookfield Zoo), Jeffrey Rogers (Yale University), Duke University Primate Center Staff, Meredith Small (Cornell University)

96   M. L. Weiss and T. R. Turner

and Ellen Merz (La Montagne des Singes). Drs Linda Smith and Ann Sodja provided many useful comments, and Dr Sodja also provided much technical support. The research was supported by NSF-BNS–8818405, NSF-BNS–9042139 and a Wayne State University Biomedical Research Support grant.

Note added in proof

Since submission of this manuscript considerable work has been done on primate DNA fingerprinting. The reader is directed to the following additional information: Martin, R. D., Dixson, A. F. and Wickings, E. J. (ed.) (1992). *Paternity in Primates: Genetic Tests and Theories: Implications of Human DNA Fingerprinting*. Basel: Karger.

References

4

Acton, R. T. and Harman, L. (1989). The use of DNA typing in cases of disputed paternity. *Promega Notes*, **22**, 1–3.
Armour, J. A. L., Patel, I., Thein, S. L., Fey, M. F. and Jeffreys, A. J. (1989a). Analysis of somatic mutations at human minisatellite loci in tumors and cell lines. *Genomics*, **4**, 328–34.
Armour, J. A. L., Wong, A., Wilson, V., Royle, N. J. and Jeffreys, A. J. (1989b). Sequences flanking the repeat arrays of human minisatellites: association with tandem and dispersed repeat elements. *Nucleic Acids Research*, **17**, 4925–35.
Avise, J. C., Bowen, B. W. and Lamb, T. (1989). DNA fingerprints from hypervariable mitochondrial genotypes. *Molecular and Biological Evolution*, **6**, 258–69.
Balazs, I., Baird, M., Clyne, M. and Meade, E. (1989). Human population genetic studies of five hypervariable DNA loci. *American Journal of Human Genetics*, **44**, 182–90.
Balazs, I., Purrello, M., Rubinstein, P., Alhadeff, B. and Siniscalco, M. (1982). Highly polymorphic DNA site D14S1 maps to the region of Burkitt lymphoma translocation and is closely linked to the heavy chain g1 immunoglobin locus. *Proceedings of the National Academy of Sciences USA*, **79**, 7395–9.
Barker, D., Green, P., Knowlton, R., Schumm, J., Lander, E., Oliphant, A., Willard, H., Akots, G., Brown, V., Gravius, T., Helms, C., Nelson, C., Parker, C., Rediker, K., Rising, M., Watt, D., Weiffenbach, B. and Donis-Keller, H. (1987). Genetic linkage map of human chromosome 7 with 63 DNA markers. *Proceedings of the National Academy of Sciences USA*, **84**, 8006–10.
Boehm, C. D., Antonarakis, S. E., Phillips, J. A., Stetten, G. and Kazazian, H. H., Jr (1983). Prenatal diagnosis using DNA polymorphisms. *New England Journal of Medicine*, **308**, 1054–8.
Boerwinkle, E., Xiong, W., Fourest, E. and Chan, L. (1989). Rapid typing of tandemly repeated hypervariable loci by the polymerase chain reaction: Application to the apolipoprotein B 3′ hypervariable region. *Proceedings of the National Academy of Sciences USA*, **86**, 212–16.

Buettner-Janusch, J., Olivier, T. J., Ober, C. L. and Chepko-Sade, C. D. (1983). Models for lineal effects in rhesus group fissions. *American Journal of Physical Anthropology*, **61**, 347–53.

Burke, T. and Bruford, M. W. (1987). DNA fingerprinting in birds. *Nature*, **327**, 149–52.

Burke, T., Davies, N. B., Bruford, M. W. and Hatchwell, B. J. (1989). Parental care and mating behavior of polyandrous dunnocks *Prunella modularis* related to paternity by DNA fingerprinting. *Nature*, **338**, 249–51.

Cann, R. L., Stoneking, M. and Wilson, A. C. (1987). Mitochondrial DNA and human evolution. *Nature*, **325**, 31–6.

Ciani, A. C., Stanyon, R., Scheffrahn, W. and Sampurno, B. (1989). Evidence of gene flow between Sulawesi Macaques. *Journal of Primatology*, **17**, 257–70.

Collick, A. and Jeffreys, A. J. (1990). Detection of novel minisatellite-specific DNA-binding protein. *Nucleic Acids Research*, **18**, 625–9.

Coppenhaver, D. H. and Oliver, T. J. (1986). Immunoglobin allotypes of Kenyan olive baboons: troop frequencies, linkage disequilibria, and comparisons with other studies. *International Journal of Primatology*, **7**, 335–50.

Cords, M. (1988). Mating systems of forest guenons: a preliminary review. In *A Primate Radiation: evolutionary biology of the African guenons*, ed. A. Gautier-Hion, F. Bourlière, J. P. Gautier and J. Kingdon, pp. 323–39. Cambridge: Cambridge University Press.

Data Acquisition and Statistical Analysis for DNA Typing Laboratories (1990). In *International Symposium on Human Identification 1989, Proceedings*. Madison: Promega Corp.

Devor, E. J. and Burgess, A. K. (1989). Short synthetic oligonucleotide repeats detect human genomic variation. *Human Biology*, **61**, 533–41.

Dixson, A. F., Hastie, N., Patel, I. and Jeffreys, A. J. (1988). DNA 'fingerprinting' of captive family groups of common marmosets (*Callithrix jacchus*). *Folia Primatologica*, **51**, 52–5.

Dracopoli, N. C., Brett, F. L., Turner, T. R. and Jolly, C. J. (1983). Patterns of genetic variability in the serum proteins of the Kenyan vervet monkey (*Cercopithecus aethiops*). *American Journal of Physical Anthropology*, **61**, 39–49.

Duvall, S. W., Bernstein, I. S. and Gordon, T. P. (1976). Paternity and status in a rhesus monkey group. *Journal of Reproduction and Fertility*, **47**, 25–31.

Eaton, G. G. (1978). Longitudinal studies of sexual behavior in the Oregon troop of Japanese macaques. In *Sex and Behavior. Status and prospectus*, ed. T. E. McGill, D. A. Dewsbury and B. D. Sachs, pp. 35–59. New York: Plenum Press.

Ely, J. and Ferrell, R. E. (1990). DNA 'fingerprints' and paternity ascertainment in chimpanzees (*Pan troglodytes*). *Zoo Biology*, **9**, 91–8.

Ewens, W. J., Spielman, R. S. and Harris, H. (1981). Estimation of genetic variation at the DNA level from restriction endonuclease data. *Proceedings of the National Academy of Sciences USA*, **78**, 3748–50.

Fooden, J. and Lanyon, S. M. (1989). Blood protein allele frequencies and phylogenetic relationships in *Macaca*: a review. *American Journal of Primatology*, **17**, 209–41.

Fossey, D. (1984). Infanticide in mountain gorillas (*Gorilla gorilla beringei*) with comparative notes on chimpanzees. In *Infanticide: comparative and evolutionary perspectives*, ed. G. Hausfater and S. D. Hrdy, pp. 217–36. New York: Aldine.

Fowler, S. J., Gill, P., Werrett, D. J. and Higgs, D. R. (1988). Individual specific DNA fingerprints from a hypervariable region probe: alpha-globin 3′ HVR. *Human Genetics*, **79**, 142–6.

Georges, M., Hilbert, P., Lequarré, A. S., Leclerc, V., Hanset, R. and Vassart, G. (1988b). Use of DNA bar codes to resolve a canine paternity dispute. *Journal of the American Veterinary Medical Association*, **193**, 1095–100.

Georges, M., Lequarré, A. S., Castelli, M., Hanset, R. and Vassart, G. (1988a). DNA fingerprinting in domestic animals using four different minisatellite probes. *Cytogenetics and Cell Genetics*, **47**, 127–31.

Gill, P., Jeffreys, A. J. and Werrett, D. J. (1985). Forensic application of DNA 'fingerprints'. *Nature*, **318**, 577–9.

Goodbourn, S. E. Y., Higgs, D. R., Clegg, J. B. and Weatherall, D. J. (1983). Molecular basis of length polymorphism in the human ζ-globin gene complex. *Proceedings of the National Academy of Sciences USA*, **80**, 5022–6.

Gouzoules, S. (1984). Primate mating systems, kin associations, and cooperative behavior: evidence for kin recognition? *Yearbook of Physical Anthropology*, **27**, 99–134.

Halpin, Z. T. (1987). Natal dispersal and the formation of new social groups in a newly established town of black-tailed prairie dogs (*Cynomys ludovicianus*). In *Mammalian dispersal patterns*, ed. B. D. Chepko-Sade and Z. T. Halpin, pp. 104–18. Chicago: University of Chicago Press.

Hausfater, G. (1975). Dominance and reproduction in baboons (*Papio cynocephalus*). *Contributions in Primatology*, **7**, 1–150.

Hoelzel, A. R. and Amos, W. (1988). DNA fingerprinting and 'scientific' whaling. *Nature*, **333**, 305.

Jarman, A. P. and Higgs, D. R. (1988). A new hypervariable marker for the human α-globin gene cluster. *American Journal of Human Genetics*, **43**, 249–56.

Jarman, A. P. and Wells, R. A. (1989). Hypervariable minisatellites: recombinators or innocent bystanders? *Trends in Genetics*, **5**, 367–71.

Jarman, A. P., Nicholls, R. D., Weatherall, D. J., Clegg, J. B. and Higgs, D. R. (1986). Molecular characterization of a hypervariable region downstream of the human α-globin gene cluster. *EMBO Journal*, **5**, 367–71.

Jeffreys, A. J. (1987). Highly variable minisatellites and DNA fingerprints. *Biochemical Society Transactions, Twenty-Third Colworth Medal Lecture*, **15**, 309–17.

Jeffreys, A. J. and Morton, D. B. (1987). DNA fingerprints of dogs and cats. *Animal Genetics*, **18**, 1–15.

Jeffreys, A. J., Brookfield, J. F. Y. and Semeonoff, R. (1985c). Positive identification of an immigration test-case using human DNA fingerprints. *Nature*, **317**, 818–19.

Jeffreys, A. J., Neumann, R. and Wilson, V. (1990). Repeat unit sequence variation in minisatellites: A novel source of DNA polymorphism for studying variation and mutation by single molecule analysis. *Cell*, **60**, 473–85.

Jeffreys, A. J., Royle, N. J., Wilson, V. and Wong, A. (1988a). Spontaneous mutation rates to new length alleles at tandem repetitive hypervariable loci in human DNA. *Nature*, **332**, 278–81.

Jeffreys, A. J., Wilson, V., Neumann, R. and Keyte, J. (1988b). Amplification of human minisatellites by the polymerase chain reaction: towards DNA fingerprinting of single cells. *Nucleic Acids Research*, **16**, 10953–71.

Jeffreys, A. J., Wilson. V. and Thein, S. L. (1985a). Hypervariable 'minisatellite' regions in human DNA. *Nature*, **314**, 67–73.

(1985b). Individual-specific 'fingerprints' of human DNA. *Nature*, **316**, 76–9.

Jeffreys, A. J., Wilson, V., Thein, S. L., Weatherall, D. J. and Ponder B. A. J. (1986). DNA 'fingerprints' and segregation analysis of multiple markers in human pedigrees. *American Journal of Human Genetics*, **39**, 11–24.

Kirby, L. T. (1990). *DNA fingerprinting: an introduction*. New York: Stockton Press.

Knott, T. J., Wallis, S. C., Pease, R. J., Powell, L. M. and Scott, J. (1986). A hypervariable region 3' to the human apolipoprotein B gene. *Nucleic Acids Research*, **14**, 9215–16.

Kocher, T. D., Thomas, W. K., Meyer, A., Edwards, S. V., Pääbo, S., Villablanca, F. X. and Wilson, A. C. (1989). Dynamics of mitochondrial DNA evolution in animals: amplification and sequencing with conserved primers. *Proceedings of the National Academy of Sciences USA*, **86**, 6196–200.

Lander, E. S. (1989). DNA fingerprinting on trial. *Nature*, **339**, 501–5.

Lander, E. S. and Botstein, D. (1987). Homozygosity mapping: a way to map human recessive traits with the DNA of inbred children. *Science*, **236**, 1567–70.

Levinson, G. and Gutman, G. A. (1987). Slipped-strand mispairing: a major mechanism for DNA sequence evolution. *Molecular and Biological Evolution*, **4**, 203–21.

Lewin, R. (1989a). DNA typing on the witness stand. *Science*, **243**, 1033–35.

(1989b). Limits to DNA fingerprinting. *Science*, **243**, 1549–51.

Ludwig, E. H., Friedl, W. and McCarthy, B. J. (1989). High-resolution analysis of a hypervariable region in the human apolipoprotein B gene. *American Journal of Human Genetics*, **45**, 458–64.

McMillan, C. and Duggelby, C. (1981). Interlineage genetic differentiation among rhesus macaques on Cayo Santiago. *American Journal of Physical Anthropology*, **56**, 305–12.

Marx, J. L. (1988). DNA fingerprinting takes the witness stand. *Science*, **240**, 1616–18.

Mech, L. D. (1987). Age, season, distance, direction, and social aspects of wolf dispersal from a Minnesota pack. In *Mammalian Dispersal Patterns*, ed. B. D. Chepko-Sade and Z. T. Halpin, pp. 55–74. Chicago: University of Chicago Press.

Melnick, D. J. and Pearl, M. C. (1987). Cercopithecines in multi-male groups: genetic diversity and population structure. In *Primate Societies*, ed. B. B.

Smuts, D. L. Cheney, R. N. Seyfarth, R. W. Wrangham and P. T. Struhsaker, pp. 121–4. Chicago: University of Chicago Press.

Melnick, D. J., Jolly, C. J. and Kidd, K. K. (1984a). The genetics of a wild population of rhesus monkeys (*Macaca mulatta*). I. Genetic variability within and between social groups. *American Journal of Physical Anthropology*, **63**, 341–60.

Melnick, D.J., Pearl, M.C. and Richard, A.F. (1984b). Male migration and inbreeding avoidance in wild rhesus monkeys. *American Journal of Primatology*, **7**, 229–43.

Nakamura, Y., Carlson, M., Krapcho, K., Kanamori, M. and White, R. (1988). New approach for isolation of VNTR markers. *American Journal of Human Genetics*, **43**, 854–9.

Nakamura, Y., Leppert, M., O'Connell, P., Wolfe, R., Holm, T., Culver, M., Martin, C., Fujimoto, E., Hoff, M., Kumlin, E. and White, R. (1987). Variable number of tandem repeat (VNTR) markers for human gene mapping. *Science*, **235**, 1616–22.

Nurnberg, P., Roewer, L., Neitzel, H., Sperling, K., Pöpperl, A., Hundrieser, J., Pöche, H., Epplen, C., Zischler, H. and Epplen, J. T. (1989). DNA fingerprinting with the oligonucleotide probe $(CAC)_5/(GTG)_5$: somatic stability and germline mutations. *Human Genetics*, **84**, 75–8.

Ober, C., Olivier, T. J. and Buettner-Janusch, J. (1978). Carbonic anhydrase heterozygosity and F_{st} distributions in Kenyan baboon troops. *American Journal of Physical Anthropology*, **48**, 95–100.

Ober, C., Olivier, T. J., Sade, D. S., Schneider, J. M., Cheverud, J. and Buettner-Janusch, J. (1984). Demographic components of gene frequency change in free-ranging macaques on Cayo Santiago. *American Journal of Physical Anthropology*, **64**, 223–31.

Olivier, T. J., Buettner-Janusch, J. and Buettner-Janusch, V. (1974). Carbonic anhydrase isoenzymes in nine troops of Kenya baboons, *Papio cynocephalus* (Linnaeus, 1766). *American Journal of Physical Anthropology*, **41**, 175–90.

Pääbo, S., Higuchi, R. G. and Wilson, A. C. (1989). Ancient DNA and the polymerase chain reaction. The emerging field of molecular archaeology. *Journal of Biological Chemistry*, **264**, 9709–12.

Pereira, M. E. and Weiss, M. L. (1991). Female mate choice, male migration, and the threat of infanticide in ringtailed lemurs. *Behavioral Ecology and Sociobiology*, **28**, 141–52.

Reilly, P. R. (1990). Individual identification by DNA analysis: points to consider. *American Journal of Human Genetics*, **46**, 631–4.

Rogers, L. L. (1987). Factors influencing dispersal in the black bear. In *Mammalian Dispersal Patterns*, ed. B. D. Chepko-Sade and Z. T. Halpin, pp. 75–84. Chicago: University of Chicago Press.

Rogaev, E. I. (1989). Two novel human DNA tandem repeat families from the hypervariable DNA probe homologous to human apolipoprotein CII-gene intron and *D. virilis* satellite. *Nucleic Acids Research*, **17**, 1246.

Rotwein, P., Yokoyama, S., Didier, D. K. and Chirgwin, J. M. (1986). Genetic analysis of the hypervariable region flanking the human insulin gene. *American Journal of Human Genetics*, **39**, 291–9.

Royle, N. J., Clarkson, R. E., Wong, Z. and Jeffreys, A. J. (1988). Clustering of hypervariable minisatellites in the proterminal regions of human autosomes. *Genomics*, **3**, 352–60.

Saiki, R. K., Scharf, S., Faloona, F., Mullis, K. B., Horn, G. T., Erlich, H. A. and Arnheim, N. (1985). Enzymatic amplification of β-globin genomic sequences and restriction site analysis for diagnosis of sickle cell anemia. *Science*, **230**, 1350–4.

Shields, W. M. (1987). Dispersal and mating systems: investigating their casual connections. In *Mammalian Dispersal Patterns*, ed. B. D. Chepko-Sade and Z. T. Halpin, pp. 3–24. Chicago: University of Chicago Press.

Smith, D. G. (1981). The association between rank and reproductive success of male rhesus monkeys. *American Journal of Primatology*, **1**, 83–90.

Smith, S. and Arber, W. (1968). Host specificity of DNA produced by *E. coli*. X. *In vitro* restriction of phage replicative form. *Proceedings of the National Academy of Sciences USA*, **59**, 1300–6.

Smuts, B. B. (1985). *Sex and Friendship in Baboons*. New York: Aldine.

Southern, E. M. (1975). Detection of specific sequences among DNA fragments separated by gel electrophoresis. *Journal of Molecular Biology*, **98**, 503–17.

Stephan, W. (1989). Tandem-repetitive noncoding DNA: forms and forces. *Molecular and Biological Evolution*, **6**, 198–212.

Stoker, N. G., Cheah, K. S. E., Griffin, J. R., Pope, F. M. and Solomon, E. (1985). A highly polymorphic region 3' to the human type II collagen gene. *Nucleic Acids Research*, **13**, 4613–22.

Strum, S. C. (1982). Agnostic dominance in male baboons. An alternative view. *International Journal of Primatology*, **3**, 175–202.

Towne, B. and Devor, E. J. (1990). Effect of storage time and temperature on DNA extracted from whole blood samples. *Human Biology*, **62**, 301–6.

Turner, T. R. (1981). Blood protein variation in a population of Ethiopian vervet monkeys (*Cercopithecus aethiops aethiops*). *American Journal of Physical Anthropology*, **55**, 225–32.

Tynan, K. M. and Hoar, D. I. (1989). Primate evolution of a human chromosome 1 hypervariable repetitive element. *Journal of Molecular Evolution*, **28**, 212–19.

Tynan, K. M., Field, L. L. and Hoar, D. I. (1990). Ethnic-specific allelic variation within the DIZ2 locus. *Human Heredity*, **40**, 1–14.

Uitterlinden, A. G., Slagboom, P. E., Knook, D. L. and Vijg, J. (1989). Two-dimensional DNA fingerprinting of human individuals. *Proceedings of the National Academy of Sciences USA*, **86**, 2742–6.

Vassart, G., Georges, M., Monsieur, R., Brocas, H., Lequarré, A. S. and Christophe, D. (1987). A sequence in M13 phage detects hypervariable minisatellites in human and animal DNA. *Science*, **235**, 683–4.

Vergnaud, G. (1989). Polymers of random short oligonucleotides detect polymorphic loci in the human genome. *Nucleic Acids Research*, **17**, 7623–30.

Vigilant, L., Pennington, R., Harpending, H., Kocher, T. D. and Wilson, A. C. (1989). Mitochondrial DNA sequences in single hairs from a Southern African population. *Proceedings of the National Academy of Sciences USA*, **86**, 9350–4.

Wainscoat, J. S., Hill, A. V. S., Boyce, A. L., Flint, J., Hernandez, W., Thein, S. L., Old, J. M., Lynch, J. R., Galusi, G., Weatherall, D. J. and Clegg, J. B. (1986). Evolutionary relationships of human populations from an analysis of nuclear DNA polymorphisms. *Nature*, **329**, 491–3.

Walter, J. R. (1987). Kin recognition in non-human primates. In *Kin recognition in animals*, ed. D. J. C. Fletcher and C. D. Michener, pp. 359–94. New York: John Wiley.

Washio, K., Misawa, S. and Ueda, S. (1989). Individual identification of non-human primates using DNA fingerprinting. *Primates*, **30**, 217–22.

Weiss, M. L. (1989). DNA fingerprints in physical anthropology. *American Journal of Human Biology*, **1**, 567–79.

Weiss, M. L., Wilson, V., Chan, C., Turner, T. and Jeffreys, A. J. (1988). Application of DNA fingerprinting probes to old world monkeys. *American Journal of Primatology*, **16**, 73–9.

Weiss, R. (1989). Doling out DNA. *Science News*, **135**, 72–4.

Weller, P., Jeffreys, A. J., Wilson, V. and Blanchetot, A. (1984). Organization of the human myoglobin gene. *EMBO Journal*, **3**, 439–46.

Wetton, J. H., Carter, R. E., Parkin, D. T. and Walters, D. (1987). Demographic study of a wild house sparrow population by DNA fingerprinting. *Nature*, **327**, 147–9.

White, R. M. and Greenwood, J. J. D. (1988). DNA fingerprinting and the law. *Modern Law Review*, **51**, 145–55.

White, R., LaLovel, J. M., Leppert, M., Lathrop, M., Nakamura, Y. and O'Connell, P. (1989). Linkage maps of human chromosomes. *Genome*, **31**, 1066–72.

Wilson, E. O. (1975). *Sociobiology, the New Synthesis*. Cambridge, Mass.: Harvard University Press.

Wong, Z., Wilson, V., Jeffreys, A. J. and Thein, S. L. (1986). Cloning a selected fragment from a human DNA 'fingerprint': isolation of an extremely poly-morphic minisatellite. *Nucleic Acids Research*, **14**, 4605–16.

Wong, Z., Wilson, V., Patel, I., Povey, S. and Jeffreys, A. J. (1987). Characterization of a panel of highly variable minisatellites cloned from human DNA. *Annals of Human Genetics*, **51**, 269–88.

Wyman, A. R. and White, R. (1980). A highly polymorphic locus in human DNA. *Proceedings of the National Academy of Sciences USA*, **77**, 6754–8.

5   Molecular genetics of globin genes and human population structure

J. FLINT, J. B. CLEGG AND A. J. BOYCE

Introduction

From a cynical standpoint, it might appear that the study of human population genetics is overburdened with theory and seriously deficient in data. It is remarkable that the most assiduous empirical workers often use their findings not so much to discredit, but certainly to question, the usefulness of the complex mathematical approximations used to account for variation in gene frequency. Perhaps an even more damaging admission would be that there is very little evidence of the relative importance of those processes most often cited to explain genetic variation in human populations, namely selection, neutral mutation and genetic drift.

Where could one turn to convince the sceptic that selection and drift do shape human population structure? The best evidence for selection comes from studies of globin variants. Investigation of the distribution of sickle cell anaemia in Africa provided the first (and still the best) example of heterozygote advantage in a human population (Allison, 1964), and studies of β-thalassaemia in Sardinia (Siniscalco et al., 1966) were for a long time the only other good piece of evidence for selection operating on human populations. It is, then, perhaps not surprising that the application of molecular biology to the field of population genetics should bear its first fruits in the cultivation of globin studies. The recent molecular evidence for the African origin of man (Wainscoat et al., 1986a) and the demonstration of natural selection acting on α-thalassaemia in the Pacific (Flint et al., 1986) appear to confirm this prediction. Moreover, because it is now relatively easy to detect molecular pathology in globin genes, and because such disorders together constitute the commonest single gene diseases of man, a wealth of epidemiological data is accumulating, whose value to population genetics is only just beginning to be exploited.

There is now a vast literature on the molecular genetics of the globin genes, fortunately, not all of it relevant to the study of population structure. There are two areas, however, which cannot be ignored: the discoveries about the relation between genotype and phenotype of

103

mutations, and the investigations about the extent and nature of variation in and around the globin genes (for reviews see Bowman, 1983; Bunn and Forget, 1986; Stamatoyannopoulos *et al.*, 1987; Stamatoyannopoulos and Nienhuis, 1987; Kazazian and Boehm, 1988; Higgs *et al.*, 1989).

Mutations of the globin genes

Haemoglobin consists of two α-like and two β-like globin subunits. They are said to be 'α-like' and 'β-like' because only the common adult globins are strictly α- and β-globins. A number of other globins are produced at different stages of development, beginning with the ζ (α-like) and ε (β-like) globins of the early embryo. The genes encoding the α-like subunits are found on chromosome 16 (in the α gene cluster) and the β-like genes on chromosome 11 (in the β gene cluster) (Deisseroth *et al.*, 1977, 1978). Consequently, although α- and β-like genes are coordinately regulated, the genes themselves are inherited independently and segregate in a Mendelian fashion. In each cluster the genes are arranged so that embryonic and fetal globin genes lie upstream (5') to the adult genes (Fig. 5.1). The α cluster encodes only two different globin polypeptides (the θ gene, also in the α cluster, is almost certainly not a globin gene: Clegg, 1987) and has two genes for each (duplicated α and ζ genes); in contrast there are four β globin-like peptides, with only one coded for by a duplicated gene (γ).

The commonest mutations to affect the globin genes, and indeed the commonest mutations in man (Livingstone, 1985), are those that reduce

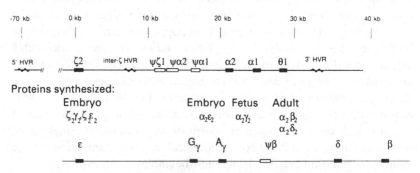

Fig. 5.1. The organization of the α- and β-globin gene clusters. Two α-like (α or ζ) and two β-like (ε, γ, δ or β) globin chains combine to form haemoglobin protein. The different combinations (or tetramers), shown in the middle of the diagram, are produced at different stages of development. The positions of three minisatellites found in the α-globin gene cluster are also shown.

or abolish the production of functional globin protein. Where α-globin is deficient, the disorder is called α-thalassaemia and where β-globin is deficient, β-thalassaemia. If no globin protein is produced it is designated an α° (or β°) thalassaemia; if the output is only reduced, the mutation is referred to as an α⁺ (or β⁺) thalassaemia. As regards the nature of these mutations, broadly speaking, they can be divided into two groups: single nucleotide changes (point mutations) and deletions. Broadly speaking again, α-thalassaemia is commonly caused by deletions (Higgs *et al.*, 1989) and β-thalassaemia and haemoglobin variants by point mutations (Antonarakis *et al.*, 1985; Kazazian and Boehm, 1988; Kazazian and Antonarakis, 1988). Much of the knowledge about point mutations, their effect on gene function and their likely origins comes from studies of β-thalassaemia; similarly, studies of the common α-thalassaemias have revealed much about how deletions arise.

Over 50 mutations have now been found to cause β-thalassaemia, the majority being single nucleotide substitutions, and small deletions or insertions (Kazazian and Boehm, 1988). These point mutations probably arise because of errors in the repair and replication of DNA and exposure to mutagens, resulting in the alteration of DNA sequence in a random fashion at a low but relatively constant rate. There is, however, some evidence that certain sequences within the β-globin gene are more mutation-sensitive than others, as Wong *et al.* (1986) argued after finding the same β-thalassaemia mutations occurring in different world populations. Point mutations can disrupt the production of stable globin at numerous points along the path from DNA to RNA to protein and these have been meticulously reported over the last decade. To give a single example, 'nonsense' mutations alter the RNA sequence so that, as occurs in the common Mediterranean position 39 mutation, a stop codon appears in the middle of the messenger RNA (Trecartin *et al.*, 1981; Orkin and Goff, 1981). Mutations have been described that affect RNA synthesis, RNA splicing, and RNA translation (Kazazian and Boehm, 1988).

Such molecular diversity of mutation was not expected, because although the clinical picture of β-thalassaemia is heterogeneous, it is not so varied as to have suggested that as many as 100 mutations might be responsible. Even more confusingly, different clinical pictures can be found with the same underlying mutation (Weatherall, 1986). The molecular characterization of these mutations shows how difficult it will be to explain their distribution within populations. Not only is it technically demanding to determine the prevalence of so many different mutations: it is also almost impossible to correlate genotype with phenotype. This makes predicting the effects of selection on populations very hard. However, as discussed later, there are now enough epidemiological data for the β-thalassaemias to begin to explain their distribution.

Two important and related ways in which DNA sequences can change are unequal crossing-over and gene conversion, mechanisms that are often invoked to explain the origin of globin gene mutations. Unequal crossing-over is responsible for the common α-thalassaemias: it can occur when homologous segments of DNA misalign. Sequencing studies have shown that the α-globin gene cluster has three blocks of homology (referred to as X, Y, Z boxes), interrupted by non-homologous DNA (Michelson and Orkin, 1983; Hess et al., 1984); recombination between misaligned Z and X boxes can give rise to DNA duplication and deletion. If two X boxes misalign and a crossover occurs (in a process referred to as reciprocal recombination) then 4.2 kb, the distance from the first to the second X box, is deleted on one chromosome and duplicated on the other, producing chromosomes with one and three α-globin genes respectively (Fig. 5.2). The deletion is referred to as a $-\alpha^{4.2}$ chromosome, the triplication as an $\alpha\alpha\alpha^{\text{anti } 4.2}$. The products of a reciprocal recombination between Z boxes are similarly referred to as a $-\alpha^{3.7}$ and $\alpha\alpha\alpha^{\text{anti } 3.7}$ (Embury et al., 1980; Higgs et al., 1980; Goossens et al., 1980; Trent et al., 1981b; Lie-Injo et al., 1981). There is an additional complication in the latter case, because the crossovers are known to occur in different places in the Z boxes, resulting in three detectably different deletions, termed $-\alpha^{3.7\text{I}}$, $-\alpha^{3.7\text{II}}$ and $-\alpha^{3.7\text{III}}$ (Higgs et al., 1984). The length of the homologous segments which have undergone misalignment and crossover (1436 bp for $-\alpha^{3.7\text{I}}$, 171 bp for $-\alpha^{3.7\text{II}}$ and 46 bp for $-\alpha^{3.7\text{III}}$) correlates with the incidence of the deletions: the smaller the homologous segment, the less often the deletion is found, an observation which may reflect an association with recombination rates. The converse is that without areas of repeated DNA, deletions occur far less readily. In the β-globin gene cluster (which has duplicated γ genes) triple (Trent et al., 1981a), single (Sukumaran et al., 1983) and even quadruple γ genes (Hill et al., 1986; Trent et al., 1986; Shimasaki and Suchi, 1986) have been found, but entire deletions of the β gene are rare. This is probably the reason why the common β-thalassaemias are attributable to point mutations, while the common α-thalassaemias are caused by deletions.

Gene conversion occurs when two sequences, which have a high degree of similarity, are made absolutely identical by a transfer of information from one DNA duplex to the other. This transfer is non-reciprocal, in contrast to the recombination processes discussed above; nevertheless, there is a relationship between the mechanisms of meiotic recombination

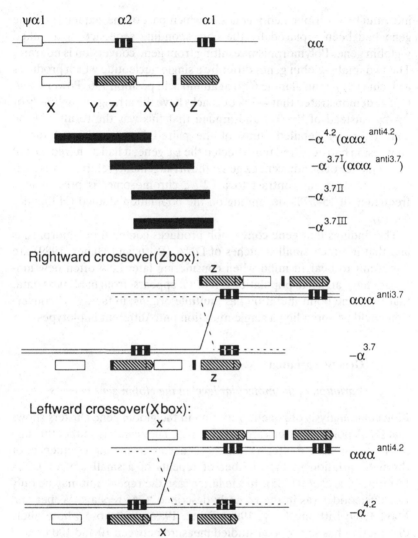

Fig. 5.2. Common deletions of the α-globin gene cluster. Three stretches of DNA are duplicated in the α cluster; they are referred to as the X, Y, and Z boxes; the Z boxes contain the α2 and α1 genes. Solid boxes indicate the approximate extent of the common α-globin gene deletions discussed in the text. The lower part of the figure shows how deletions and duplications are thought to arise by crossing over the misaligned chromosomes.

and gene conversion, not yet fully understood, as is revealed by the fact that gene conversion occurs at between 30 and 70% of all crossings over (Orr-Weaver and Szostak, 1985). Slightom *et al.* (1980) provided the first evidence that conversion operated on human genes when they found an individual with triplicated γ genes in which part of the sequence of one gene had been replaced by the corresponding sequence of a linked γ-globin gene. Polymorphism resulting from gene conversion is not rare. The two fetal γ-globin genes differ by a single nucleotide which produces a glycine ($^G\gamma$) or an alanine ($^A\gamma$) at amino acid position 136. Powers *et al.* (1984) demonstrated that 4–8% of Blacks have the arrangement $^G\gamma$–$^G\gamma$ or $^A\gamma$–$^A\gamma$ (instead of $^G\gamma$–$^A\gamma$) and argued that this was the result of gene conversion. A detailed study of the paired ζ genes showed that a conversion had occurred that affected the $\psi\zeta$ gene; it had acquired part of the sequence of an adjacent $\zeta2$ gene; the arrangement, referred to as a $\zeta2$-$\zeta1$ chromosome in contrast to a $\zeta2$-$\psi\zeta$ chromosome, is present at a frequency of 15–57% depending on the population studied (Hill *et al.*, 1985b).

The findings that gene conversion produces common polymorphisms and that it affects small stretches of DNA (Borts and Haber, 1989) are important to bear in mind when considering later how often new mutations have arisen in populations. It often appears, from haplotype data, that the same point mutation has multiple origins; in fact, gene conversion could be spreading a single mutation onto different haplotypes.

Genetic variation

Variation at the molecular level in the globin gene clusters

Molecular analysis of genetic variation in the globin gene clusters shows that DNA polymorphisms are of two types. They can arise either through point mutations that alter a restriction enzyme recognition sequence, or through variation in the number of repeats of a small unit of DNA (Antonarakis *et al.*, 1985). In the latter case the repeat unit may be only two nucleotides (as in the recently described 'CA' repeats: Weber and May, 1989; Litt and Luty, 1989; Tautz, 1989), but those whose allele distribution has so far been studied measure between 10 and 100 bp and are referred to in the literature as 'minisatellites', 'hypervariable regions' (HVRs) or 'variable number tandem repeats' (VNTRs) (Jeffreys *et al.*, 1985; Nakamura *et al.*, 1987). They can be detected by any enzyme that cuts reasonably close (say within 5 kb) to the variable region. Point mutations require specific enzymes for their detection, and are conventionally named after the enzymes used. It follows that many single nucleotide changes are undetected, though the sequencing of numerous β-globin genes has indicated the extent of this hidden variation (Savatier *et al.*, 1987).

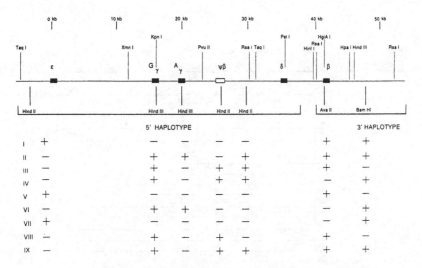

Fig. 5.3. Polymorphic restriction enzyme sites in the β-globin gene cluster. Beneath the gene map are the seven sites most frequently reported in the literature combined into a haplotype. They are divided into two sub-haplotypes by an area of increased recombination between the δ- and β-globin genes. Conventionally the 5' end of the DNA molecule is put on the left of diagrams (also called 'upstream'). Consequently, the five upstream sites of the sub-haplotype are termed a 5' haplotype and the two downstream sites a 3' haplotype.

The commonly used restriction site polymorphisms and minisatellites in the globin gene clusters are shown in Figs 5.3 and 5.4. Restriction site polymorphisms in both clusters have one important, indeed invaluable, characteristic: they are close enough to be in linkage disequilibrium with each other; in other words, they do not segregate independently. This explains why the concept of a haplotype is so useful in population studies. The idea itself is straightforward: take, for instance, the five 5' polymorphisms in the β cluster, from the ε Hind II to the $\psi\beta$ Hind II. The presence of a site produces a smaller length of DNA and is denoted by $+$, the site's absence by $-$. If all five sites in an individual were analysed it might be found that both chromosomes have the Hind II site at the ε gene, while in the γ genes all four Hind III sites are missing and at the $\psi\beta$ Hind II site one chromosome is $+$ and one $-$. One would then conclude that one chromosome has the combination of polymorphic sites $+----$ and the other $+---+$. The combination of sites is called a haplotype (and can of course consist of fewer or more than five sites). Because the sites are so close together on the chromosome, they will be transmitted together and there is a very low chance that recombination will disrupt the haplotype

	Xbal	Sacl	Bgll	SML	PZ/Z	Accl	Rsal	Pstl	Pstl
Ia	+	+	-	M	PZ	+	+	-	-
Id	-	+	-	L	PZ	+	+	-	-
IIa	-	+	-	L	PZ	+	-	-	-
IIc	+	+	-	M	PZ	+	-	-	-
IId	-	-	-	M	PZ	+	-	-	-
IIg	+	-	-	M	PZ	+	-	-	-
IIIa	-	-	+	M	Z	-	-	-	-
IIIb	+	-	+	M	Z	-	-	-	-
IIIf	-	-	-	M	Z	-	-	-	-
IVa	+	-	-	S	PZ	+	-	+	+
Va	-	+	-	L	PZ	+	-	-	+
Vc	+	-	-	S	PZ	+	-	-	+
VIIa	-	+	-	M	Z	+	-	-	-

Fig. 5.4. The seven restriction enzyme sites used in the construction of the α-globin cluster haplotype are shown in the box below the gene map. Two other polymorphisms are also included in the haplotype: firstly the inter-ζ hypervariable region (HVR) (or minisatellite) where the commonly detected alleles are reported as small (S), medium (M), or large (L). Secondly, as explained in the text, the $\psi\zeta1$ gene has a variant form ($\zeta1$) which is detected by a length difference in its first intron's hypervariable region. The two forms are designated in the haplotype as PZ (pseudozeta) and Z (zeta).

Infrequent polymorphisms are shown above the gene map. At either end of the cluster are two minisatellites: the 5' and 3' hypervariable regions. Different 5' and 3' HVR alleles can be associated with each haplotype; this of course dramatically increases the possible number of observed haplotypes. The population distribution of these HVR alleles has not yet been fully explored and is not here included in the haplotypes.

The box shows the structure of ten commonly found haplotypes; the frequencies of these haplotypes in different world populations are shown in Table 5.2.

from generation to generation (i.e. the sites are in linkage disequilibrium). Therefore each haplotype exists as a stable allele in the population, and can be treated as such in analysis of population structure.

β-globin haplotypes are often reported in the literature using the nomenclature introduced by Orkin and Kazazian (1984), who were the first to explore β haplotype distributions systematically. Their system is based on their initial survey of chromosomes in the Mediterranean and is arbitrary (Orkin et al., 1982). They described nine seven-site haplotypes, numbered I to IX (see Fig. 5.3). Higgs et al. (1986) classified the α-globin haplotypes by their four 3' sites and by the presence (or absence) of a $\psi\zeta$ gene; this system defines five major haplotypes (see Fig. 5.4). They are subdivided by the inclusion of three 5' sites and a minisatellite between the ζ genes which is reported as small (S), medium (M), or large (L) allele. A total of 29 haplotypes was thus originally identified. The greater complexity of the α-globin haplotypes makes it difficult, in the absence of pedigree data, to determine complete haplotypes. Without adequate family data only homozygotes and heterozygotes that differ at one site can be confidently assigned the correct haplotypes.

α and β haplotypes differ in two respects. In the first place only the α cluster has minisatellites. There are five reported; one between the two ζ genes (the minisatellite mentioned above) (Goodbourn *et al.*, 1983), in the first intron of both ζ genes, in the second intron of the same gene (Proudfoot *et al.*, 1982), 3' to the θ gene (Jarman and Higgs, 1988) and about 70 kb 5' to the $\zeta2$ gene (Jarman *et al.*, 1986). Variation is most marked at the latter two loci, a heterozygosity of 90% having been reported at the 3'HVR (Jarman *et al.*, 1986). Secondly, recombination frequency is not constant across the β cluster, while it appears to be so across the α cluster. Antonarakis *et al.* (1982b) showed that polymorphic sites were not randomly associated across the β-globin cluster and proposed that this occurred in two distinct segments with increased recombination between the two. They described a five-site 5' haplotype and a two-site 3' haplotype (Fig. 5.3), and suggested that there was an area of increased recombination between the δ- and β-globin genes. Support for this theory came both from population surveys, which reveal that 75% of the recombination in the 63 kb of the β-globin cluster occurs within this region (Chakravarti *et al.*, 1986), and from *in vitro* study of the recombinational activity of this area. Insertion into yeast of 1.9 kb of DNA situated 2 kb upstream of the β-globin gene increased recombination fourfold compared with controls (Treco *et al.*, 1985). Moreover, occasional families have been found in which recombination between the 5' and 3' haplotypes has been directly observed (Gerhard *et al.*, 1984; Old *et al.*, 1986; Camaschella *et al.*, 1988b).

While an area of increased recombination frequency doubtless exists in the β-globin cluster, initial estimates of the degree of recombination, put as high as 30 times the normal (Chakravarti *et al.*, 1986) are vitiated by very large standard errors (Weir and Hill, 1986; Hill and Weir, 1988). It is not yet possible to quantify the effect of this recombination 'hotspot', but the important result for population studies is that there is less linkage disequilibrium between a 5' and a 3' haplotype than between the sites that constitute the sub-haplotypes.

How much variation is present in the gene clusters?

When molecular biologists first started to sequence β-globin genes in order to determine the mutations responsible for β-thalassaemia, there was concern as to whether the degree of polymorphism between individuals might be so great that it would be impossible to decide which differences were normal variants and which were responsible for the thalassaemias. This fear was unfounded. The sequences of normal β-globin genes were remarkably invariant, and in fact to date only seven different normal β-globin genes are known, the differences being caused by variation at ten polymorphic sites (Lawn *et al.*, 1980; Orkin *et al.*, 1982; Poncz *et al.*, 1983; Kimura *et al.*, 1983; Antonarakis *et al.*, 1985; Savatier *et al.*, 1987). Looking at the number of β-globin haplotypes, a similar picture emerges. Taking the five sites of the 5′ β-globin haplotype, there are in theory 2^5 (32) possible haplotypes. To date, out of 5000 chromosomes typed, only 24 different β-globin haplotypes have been found (Table 5.1). Even more significantly, four haplotypes make up more than 99% of the total. Even with more than 20 alleles the total heterozygosity at this locus is only 60%. These observations apply to the worldwide total, making no distinction between racial groups.

The distribution of α-globin haplotypes is similar. Again there is marked linkage disequilibrium. Using all possible combinations of alleles in the cluster, a total of 768 haplotypes is theoretically possible, although only 29 have so far been found (Higgs *et al.*, 1986). Again a small number of haplotypes predominate, though not enough data have yet been collected to estimate accurately the haplotypes' gene frequencies (Table 5.2). Tight linkage has been observed between certain sites and haplotypes (making it possible to determine the frequencies of some haplotypes relatively easily). For instance, only the type Ia has an Rsa site and outside Africa only group III haplotypes have the ζI rather than the $\psi\zeta1$ 3′ ζ-globin gene (Higgs *et al.*, 1986).

There are a number of qualifications that need to be made before deciding what this very skewed distribution of haplotypes implies. First, is this a true estimate of the amount of variation in the clusters? May it not be that there is far more, and that this is not detected by the enzymes used? Whenever a '+' is recorded in the haplotypes, indicating the presence of a restriction site, we know the exact sequence at that point. If two chromosomes share a '+' they are identical at the site. The same is not true of a '−'. All that this tells us is that one 6 base pair sequence is not present (assuming the restriction enzyme recognizes a 6 bp sequence). We also do not know anything about the DNA between the restriction sites. How many additional differences may there be? A simple way to assess this is to look for other polymorphisms in the cluster. Chakravarti *et al.* (1986) estimated the linkage between two more sites within the 5′ β-globin haplotype. They found complete association between these sites and the second Hind II site in Mediterranean and Asian populations (though this did not hold for American Blacks). Similarly, studies using larger numbers of 3′ β-globin polymorphisms do not dramatically increase the number of haplotypes found using just the Ava II and Bam HI sites (three). For instance, using six sites, Kazazian *et al.* (1986a) found five haplotypes in 78 chromosomes, out of which one occurred only once and another in only three cases (so three haplotypes still accounted for 99% of the total). Certainly there is more polymorphism than the seven-site haplotype detects, but it is not so much as to invalidate conclusions drawn from using only seven sites in the β-globin haplotype.

The analysis of genetic variation in human populations

Although the issue is still much discussed, there is general agreement that the distribution of genetic variants cannot be entirely explained by selection: the processes of genetic drift and founder effect have also to be taken into account. Data from globin gene studies illustrate the contribution of both neutral and selective forces in shaping the genetic structure of populations. They show particularly well how the vicissitudes of ancient populations still leave traces on contemporary population structure through genetic drift and founder effects; they also provide evidence for selection acting on human populations and that malaria is responsible for the high frequencies of a number of haemoglobinopathies. These two determinants of genetic diversity are discussed separately below.

Genetic diversity: the contribution of population movements

A population bottleneck in the human settlement of Eurasia

As was noted above, examining the frequencies of both α- and β-globin haplotypes indicates that in all populations both loci show a similar pattern. There are two or three common haplotypes and a great number of rarer ones. However, not all world populations share the same common haplotype. This was first noticed for the 5′ β haplotypes, where the most outstanding difference was observed between West African and other world populations (Table 5.1) (Wainscoat *et al.*, 1986a). West Africans possess two haplotypes, occurring at 60% and 20%, that are at frequencies of less than 1% elsewhere in the world. In contrast, Eurasians

Table 5.1. 5' β-globin gene haplotypes for different populations. Both normal (N) and thalassaemic (T) chromosome haplotypes are given

	Europe														
	Italy		Iberia	Germany	Greece		Britain	Lebanon	Sardinia		Cyprus		Turkey	Total Europe	
Haplotype	N	T	T	N	N	T	N	T	N	T	N	T	T	N	T
1 + − − − −	61.8	43.6	43.8	43.8	59.3	69.4	43.2	64.7	58.7	18.2	70.0	90.9	57.5	61.0	49.0
2 − + + − +	10.9	32.9	46.6	6.3	18.5	14.2	13.5	13.7	21.5	66.7	7.5		6.0	13.1	29.6
3 − + − + +	25.3	9.1		31.3	8.6	5.2	40.5	7.8	13.2	6.1	20.0	1.3	18.7	21.2	8.3
4 − − − + +					1.2									0.3	
5 − − − − +				6.3	3.7	3.0		7.8						1.0	1.0
6 + + − − −	1.0	13.6	8.2	12.5	6.2	8.2	2.7	7.8	3.3	3.8	2.5	7.8	16.4	2.7	11.5
7 − + − − −	0.3	0.2	1.4		1.2			5.9	3.3	5.3				0.3	0.2
8 + − − + +	0.3													0.1	
9 − − − − −	0.3													0.1	
10 + + − − + +					1.2									0.1	
11 + − − − − +															
12 − − − + +															
13 + + + − +															
14 + + − − −															
15 + + + − −		0.2											1.5		0.1
16 − + − + +		0.5													0.2
17 − + + + +															
18 + − − + −															
19 − − − + +															
20 − − + + −															
21 + − + − −															
22 − − + − −															
Totals	304	660	73	16	81	134	37	51	121	132	120	77	134	679	1259

Table 5.1. (*cont.*)

	ASIA											
	China		Thailand		Cambodia	Japan	India		Indonesia		Total Asia	
Haplotype	N	T	N	T	N	N	N	T	N	T	N	T
1 + − − − − −	79.7	80.6	75.7	77.9	72.3	66.7	53.7	71.9	90.0	72.9	68.3	77.8
2 − + + − +	10.0	12.8		5.1	4.3	11.1	10.3	5.3	1.7		7.8	9.1
3 − + − + +	6.9	0.9	17.3	4.6	19.1	11.1	25.6	21.0	3.3	24.3	17.4	7.4
4 − − − +		0.4						1.8				0.2
5 − + − − +			1.2	2.1	4.3		4.7				2.4	0.7
6 − + + − −				1.0			1.2		1.7	1.4	0.5	0.2
7 − + − − −				0.5			0.2		3.3	1.4	0.2	0.2
8 + − − − + +	0.9			1.5			0.2				0.5	0.3
9 − − − − −		3.4		2.6			0.5				0.2	2.3
10 + + + − +							1.7				0.7	
11 + − − − − +				2.1								0.3
12 − − + +	2.2	0.3	2.1				0.5				1.2	0.2
13 + + + + + −	0.4						0.5				0.3	
14 + + + − + −						11.1	0.7				0.5	
15 + + + −												
16 − + − + −				1.0							0.1	0.2
17 − + + + +			0.4									
18 + − − + −		0.1		0.5								0.2
19 − − + + −		1.3										0.7
20 − − − + −		0.1										0.1
21 + − + − −				1.0								0.2
22 − − + − −												
Totals	231	682	243	195	47	18	406	281	60	70	1005	1228

Table 5.1. (*cont.*)

| Haplotype | AFRICA | | | | | | | | OCEANIA | | | | WORLD | |
| | West Africa | North Africa | | South Africa | US Blacks | | Total Africa | | Melanesia | Micronesia | Polynesia | Total Oceania | | |
	N	N	T	N	N	T	N	T	N	N	N	N	N	T
1 + − − − −	4.8	56.5	44.8	8.0	21.4	10.4	20.0	35.1	66.1	93.6	90.7	80.6	64.5	60.4
2 − + + − +	2.9	8.7	16.5	12.6	7.1	58.4	7.7	28.4	2.2		1.1	0.9	7.2	20.4
3 − + − + +	12.4	23.2	24.7	14.9	21.4	11.7	17.0	21.0	15.6		4.5	7.9	15.7	9.2
4 − − − − +	55.2	4.3	3.1	34.5	26.2	9.1	34.0	4.8	1.1			0.3	3.9	0.6
5 − + − − +	17.1			25.3	14.3	10.4	15.3	3.0	3.2		1.5	3.6	3.8	1.1
6 − + + − −		2.9	6.2		2.4		1.0	4.4		2.0		0.5	1.1	5.8
7 − + − − −	1.9		0.5	2.3	2.4		1.7	0.4	0.5		0.7	0.4	0.4	0.2
8 + − − + +									6.5			1.9	0.7	0.1
9 − − − − −		4.3	3.6				1.0	2.6	1.6	1.0	0.4	0.5	0.4	1.3
10 + + − − +													0.3	
11 + − − − +	2.9			1.1			1.3			0.5	0.4	0.4	0.3	0.1
12 − − − + +				1.1			0.3			1.5		0.4	0.6	0.1
13 + + + + −	1.9						0.6						0.2	
14 + + − − −													0.4	
15 + + + − −			0.5		4.8			0.4				0.9		
16 − + − + −							0.7						0.1	0.1
17 − + + + +									2.7	1.5	0.7	1.3	0.4	0.2

		105	69	194	87	42	77	300	271	186	204	269	749	2735	2760
18	+ – – + –														0.1
19	– – + + –														0.3
20	– – – + –														0.1
21	+ – + – –												0.3	0.1	0.1
22	– – + – –							0.5							0.1
Totals		105	69	194	87	42	77	300	271	186	204	269	749	2735	2760

Data for this and Tables 5.5 and 5.6 are from the following references: Akar et al., 1987; Amselem et al., 1988; Antonarakis 1982a,b, 1984b, 1988a; Athanassiadou et al., 1987; Aulehla-Scholz et al., 1990; Camaschella et al., 1988a; Carestia et al., 1987; Chan et al., 1986, 1987; Chehab et al., 1987; Cheng et al., 1984; Chibani et al., 1988; Coutinho-Gomes et al., 1988; Del Senno et al., 1985; Diaz-Chico et al., 1988; Di Marzo et al., 1988; Fucharoen et al., 1989; Fukumaki et al., 1988; Giampaolo et al., 1984; Hararo et al., 1985; Huang et al., 1985, 1986, 1990; Hundrieser et al., 1988; Kaplan et al., 1990; Kazazian et al., 1984a,b, 1986a, 1987; Laig et al., 1989; Lie-Injo et al., 1989; Liu et al., 1988; Long et al., 1988; Lynch et al., 1988; Maggio et al., 1986, 1988; Millard et al., 1987; Monteiro et al., 1989; Oehme et al., 1985; Old et al., 1984; Orkin et al., 1982; Ottolenghi and Carestia, 1986; Pirastu et al., 1987, 1988; Ramsay and Jenkins 1987; Rosatelli et al., 1985, 1987, 1988; Rouabhi et al., 1988; Sampietro et al., 1988; Shimizu 1987; Sozuoz et al., 1988; Thein et al., 1984, 1988; Wainscoat et al., 1983a,c, 1986a,b; Wong et al., 1986; Yenchitsomanus et al., 1988; Yongvanit et al., 1989; Zeng and Huang 1987; Zhang et al., 1988.

Table 5.2. *Gene frequencies of α-globin haplotypes in different world populations*

Haplotype	Britain	Mediter-ranean	Asian Indian	Saudi Arabia	Papua New Guinea	Island Melanesia	Poly-nesia	Micro-nesia	SEAsia	Jamaica	Nigeria	SA San	SA Black
I a	50	51	44	38	0	?	33	28	25	23	?	0	6
I d	0	0	0	0	0	0	0	0	4	0	0	0	0
II a	25	18	+	38	0	0	6	12	31	0	0	0	0
II c	0	+	+	0	0	?	2	12	6	?	?	11	3
II d/e	0	+	+	0	0	?	21	15	19	?	0	0	3
II g	0	?	?	?	0	0	0	1	2	+	?	57	11
III a	?	+	+	+	45	+	16	13	10	?	?	0	3
III b	+	?	+	?	18	33	0	1	0	+	0	5	3
III f	0	?	?	?	0	0	0	0	0	23	27	0	17
IV a	0	0	0	0	+	58	17	7	0	?	0	0	3
V c	0	0	0	0	+	?	2	3	0	?	0	11	6
VII a	?	0	?	0	0	?	0	0	0	23	?	0	0
Number of chromosomes	32	62	62	14	60	18	586	191	160	38	28	37	36

Because of the complexity of the α haplotype it is possible to work out full haplotypes only on samples with adequate family data; otherwise frequencies have to be deduced from the frequencies of homozygotes, assuming the alleles are in Hardy–Weinberg equilibrium. Complete α haplotype data are only available for some populations in Southeast Asia and the Pacific (Hertzberg *et al.*, 1988; O'Shaughnessy *et al.*, 1990) and in Africa (Ramsay and Jenkins 1988). In this table frequencies derived from numbers of observed homozygotes are taken from Higgs *et al.* (1986) except for Southeast Asia and the Pacific, and for the San and South African blacks. '+' indicates that the haplotype has been observed but its frequency cannot be estimated from the available data; the frequency is however very likely to be low. A '?' indicates that a particular haplotype may be present but this cannot be confirmed from the data available.

have three haplotypes with frequencies of 65%, 19% and 10% (that is, in total 94% of all Eurasian haplotypes) whose frequencies in West Africans are 5%, 10% and 2% (or one fifth of the total). No difference of this magnitude emerges from breaking down the Eurasian populations; the same three haplotypes remain the commonest in each subdivision, although the populations vary, notably in Southeast Asia where the $+----$ haplotype comprises 80% of the total; it is only 58% in Europe and India (Table 5.1).

What can be made of these differences? The data are taken from normal (β^A) chromosomes, supposedly therefore selectively neutral (or at least open to the same degree of selective pressure). Consequently one explanation offered is that there is a primary division between Africans and Eurasians, the haplotype frequency difference arising after a population bottleneck (Wainscoat *et al.*, 1986a; Jones and Rouhani, 1986a). If this interpretation is correct, that is, if genetic divergence between the Africans and other world populations is the result of a population bottleneck, then the haplotype data can be used in favour of the 'Out of Africa' hypothesis of *Homo sapiens* evolution (Howells, 1976; Jones and Rouhani, 1986a). It should be said immediately that the DNA data have not resolved the debate over human origins; as with many aspects of the current revolution that has arisen out of the new molecular techniques, it is a case of the wealth of new data casting unexpected light into corners far removed from those usually illuminated by molecular biology.

In brief, one way of polarizing, for heuristic reasons, the arguments about human origins is to contrast multi-regional and single origin theories (Howells, 1976; Smith and Spencer, 1984; Stringer and Andrews, 1988; Mellars and Stringer, 1989). The difficulty presented by fossil data is that before the emergence of undeniably modern man, skeletal remains from different parts of the world show considerable morphological variation. The single origin model presumes that outside Africa this variation represents various dead ends of early hominid evolution; our true ancestors differentiated into recognizably modern humans in Africa and then replaced all the possible prototypes elsewhere in the world after migrating 'Out of Africa' about 200 000 years ago (Jones, 1986; Cann *et al.*, 1987). The multi-regional alternative to this theory argues not that all over the world modern man evolved independently, but that there was sufficient gene flow to prevent speciation and allow for a more gradual, but worldwide, differentiation into *Homo sapiens* over the last million years (Wolpoff, 1989). This theory therefore describes geographically extended phyletic speciation.

Fitting the β haplotype data into these theories, it is clear that the bottleneck, assumed to account for the African–Eurasian split, could be the result of the migration of a small number of modern humans from Africa into Eurasia (Wainscoat *et al.*, 1986a). Critics of this view have not been slow to point out that the bottleneck could have occurred in the other direction, from Europe to Africa. The β-globin data are in fact compatible with an origin in Africa, Eurasia, or both continents (Jones and Rouhani, 1986a,b; van Valen, 1986; Edwards, 1986; Giles and Ambrose, 1986; Diamond and Rotter, 1987). Support for the first option is argued from one interpretation of the fossil evidence, but the inference from the DNA data that a bottleneck is attributable to population movement may also be incorrect. There is now little doubt that the sickle-cell gene (β^S) confers selective advantage in malarious regions (see below). Consequently, in malarious areas, haplotypes on sickle gene chromosomes will increase in frequency while haplotypes on normal (β^A) chromosomes will be relatively reduced in number, experiencing, in effect, a bottleneck. We know that recombination can occur between haplotypes (Gerhard *et al.*, 1984; Old *et al.*, 1986; Camaschella *et al.*, 1988a), but to what extent this process may have reshaped gene frequencies of β^A chromosomes is a matter for speculation. More sophisticated analysis may be required to take into account such factors as selection, recombination and gene conversion; one attempt to re-assess the β-globin data along these lines still comes out in favour of an African origin (Long *et al.*, 1990).

There are additional genetic data in favour of the African origin theory. Recent work on α-globin haplotypes and surveys of polymorphisms from throughout the genome also show that the African haplotype distribution differs from the Eurasian (Bowcock *et al.*, 1987; Cavalli-Sforza *et al.*, 1988; Wainscoat *et al.*, 1989). There are also a number of African-specific polymorphisms discovered elsewhere in the genome (Chakravarti *et al.*, 1984; Anagnou *et al.*, 1984; Bell *et al.*, 1984; Lucotte *et al.*, 1989). A survey of 120 alleles in 42 populations showed a primary split between Africa and Eurasia, and this division was recapitulated in a taxonomic classification of language phyla (Cavalli-Sforza *et al.*, 1988). If Africa was the ancestral home of *Homo sapiens*, with Eurasia being relatively recently colonized by a small founder population, then one prediction is that the present African population would be more genetically heterogeneous than the Eurasians. There is some evidence for this from both mitochondria (mt) DNA (Cann *et al.*, 1987) and protein and antigen polymorphisms (Nei and Roychoudhury, 1982). However, the question will only be finally resolved with further multi-disciplinary investigation of anthropological and genetic data.

Haemoglobinopathies in the Pacific

In some ways it is surprising that such a broad overview as the sampling of various world populations at a single locus presents any identifiable pattern; it might be thought that the complexities of migration, differing population sizes and mating patterns would confuse such gross comparisons. How much more, then, can be expected to be learnt from molecular mapping of small populations; a micro-epidemiological in place of a macro-epidemiological study. Investigations of this kind have been carried out in order to describe the population structure of Pacific islanders.

The Pacific is traditionally divided into three areas, roughly in accordance with ethnic divisions: Polynesia, occupied by relatively fair skinned, Austronesian speakers; Micronesia, whose inhabitants resemble the Polynesians and speak related languages; and Melanesia, which is occupied by a very diverse collection of peoples, speaking both Austronesian and Papuan languages (the latter group consisting in New Guinea of more than 1000 different languages) (Wurm, 1982), but who are strikingly different physically from Polynesians and Micronesians (Howells, 1973).

Studies of globin genes and associated DNA polymorphisms in the Pacific have revealed an extensive number of haemoglobinopathies (Hill *et al.*, 1989a). The high incidence of anaemia in Melanesia found by earlier haematological surveys of the region and thought to be caused by dietary insufficiencies and other environmental causes, turned out to be due to a high incidence of α-thalassaemia (Bowden *et al.*, 1985). Moreover, this α-thalassaemia is heterogeneous at the molecular level and previously undescribed deletions were discovered. Chromosomes bearing rearrangements of other globin genes also came to light. Triple γ genes were first found in Vanuatu (Trent *et al.*, 1981a); triple ζs occur in Polynesia (Trent *et al.*, 1986, 1988; Hill *et al.*, 1987). Even quadruple γ genes have been reported (Hill *et al.*, 1986).

What is the reason for the high frequencies of α-thalassaemia and the wealth of other globin gene variants? The ecological diversity of the Southwest Pacific and the physical and cultural difference of its human inhabitants almost *a priori* precludes a single explanation, but molecular analysis of the globin gene variants has shown how it is possible to account for some of the features of the genetic constitution of the Pacific islanders in terms of genetic drift and population movements.

Evidence of genetic drift in the Pacific

Work on protein and antigen polymorphisms has led to the description of New Guinea as a set of genetic isolates, where small population size and little migration allowed the gene frequencies to vary considerably over relatively small geographical distances (Giles *et al.*, 1970; Imaizumi and Morton, 1970; Simmons and Booth, 1971). Consequently, a reasonable guess would be that similar processes explained the epidemiology of globin gene variants in the Pacific. Published accounts bear this out for Polynesia and Micronesia, where the frequency of each variant shows a random distribution from island to island, and each variant can be accounted for by the occurrence of a single mutational event (Trent *et al.*, 1986, 1988; Hill *et al.*, 1987; O'Shaughnessy *et al.*, 1990).

A good example is the distribution of α-thalassaemia in Polynesia and Micronesia, which has a gene frequency varying from 0.1 to 15% (Hill *et al.*, 1987; Trent *et al.*, 1988; Hertzberg *et al.*, 1988; O'Shaughnessy *et al.*, 1990). The evidence that the α-thalassaemia has a single origin comes from molecular characterization of the deletion; it is entirely of the $-\alpha^{3.7III}$ variety and occurs on a single haplotype (IIIa) (Hill *et al.*, 1985a; Trent *et al.*, 1988; O'Shaughnessy *et al.*, 1990). Although this haplotype is relatively common in Oceania, it is rare elsewhere in the world (Table 5.2) (Higgs *et al.*, 1986), implying that the mutation arose locally. Moreover, there are reasons to believe that the probability of its arising more than once is small (compared with the other α-thalassaemia deletions found in the area) because the recombinational event presumed to have produced it must have involved a much shorter stretch of homologous DNA than the alternative deletions (Higgs *et al.*, 1984). All α-thalassaemia in Polynesia is therefore very likely to be descended from a single mutation and to have reached its present high frequencies because of genetic drift and founder effects. Similar reasoning can explain the distribution of the other variants. The triple ζ, triple α and single ζ chromosomes each occur on a single haplotype and have distributions restricted to various parts of the Pacific (Lie-Injo *et al.*, 1985; Hill *et al.*, 1987). Exceptions to this pattern are the single and triple γ variants which occur on two haplotypes each (suggesting two origins for each) (Hill *et al.*, 1986); however, their distribution again shows no cline. The explanation for the distribution of the globin variants is therefore partly to be found in the genetically isolated nature of Pacific populations.

Population movements in the Pacific

Reconstructions of Pacific history based on the existing archaeological record, the distribution and relationships between languages, and various anthropometric and genetic variables (including skeletal resemblances as well as protein and antigen polymorphisms) date the first appearance of man in Oceania at around 40 000 years ago in Papua New Guinea (Groube *et al.*, 1986), almost certainly coming from Southeast Asia. These first colonists, who had reached Australia by 33 000 years ago (White and O'Connell, 1982), are believed to be the ancestors of the Melanesians, but several later waves of immigrants have considerably complicated the picture, both genetically and linguistically (Wurm, 1983).

Little is known about the colonization of island Melanesia (i.e. the Melanesian islands excluding New Guinea). In the north, archaeological sites date the occupation of New Britain to about 30 000 years ago (Allen *et al.*, 1988); in the southeast (in Vanuatu) the record stretches back only a few thousand years (White and O'Connell, 1982).

Perhaps the most vexing issue in Oceanic history revolves around the question of Polynesian origins. Present opinion agrees that they came from Southeast Asia, arriving in Fiji, Samoa and Tonga between 3000 and 4000 years ago and the Marquesas Islands by about AD 200. The remainder of Polynesia was settled by AD 1000 (Brockway, 1983; Kirch, 1986; Bellwood, 1989). How a low-technology society managed to colonize an ocean as vast as the Pacific in so short a time has been a puzzle for decades. There has been much speculation about the origin of these unusual seafarers, though whether the discovery of their homeland will in fact tell us much about how the Polynesians settled in Polynesia is a moot point; on the other hand an understanding of the route they took in reaching Fiji, Tonga and Samoa, and their duration of stay in places along the way, might do so.

The third group in Oceania are the Micronesians. The physical simi-larities between Micronesians and Polynesians were noted more than 50 years ago (Buck, 1938), implying some relationship between the two peoples. Linguistically, west Micronesia is related to Indonesia and the Philippines, while east Micronesia has languages that could be derived from northern island Melanesia (Clarke, 1979; Pawley and Green, 1984). The absence of Lapita pottery, a ceramic marker of Polynesian presence, from Micronesia could also be interpreted as evidence against an affinity between Polynesia and Micronesia, but the issue is by no means resolved (Green, 1979; Spriggs, 1984).

The DNA data have been most useful in the description of Polynesian affinities; the molecular characterization of just two markers has already shown that there must have been some genetic admixture between Melanesians and Polynesians. It demonstrates that one pathway taken by the Polynesian colonists was through Vanuatu, and suggests that a second route may have been through Micronesia (Hill *et al.*, 1985a: O'Shaughnessy *et al.*, 1990). The first of these markers is the $-\alpha^{3.7III}$ deletion. It was mentioned above that this mutation probably had a single origin and is found only in the Pacific, implying that it arose there.

There is a good reason for suspecting the deletion to have arisen in Melanesia and indeed for locating it somewhere between New Britain and northern Vanuatu, because this is the area of distribution of HbJ^{Tongariki}. This globin gene variant occurs solely in association with the $-\alpha^{3.7III}$ deletion (whereas the converse does not hold). The unusual conjunction of the two mutations suggests that the $-\alpha^{3.7III}$ deletion is the older of the pair (because it is the more widely distributed) and must have been present at relatively high frequencies somewhere in northern island Melanesia when HbJ^{Tongariki} first appeared. This argument does not rule out the possibility that the deletion arose outside Vanuatu and was then elevated by selection to its present frequency, but makes such a reconstruction less plausible. Instead it seems likely that the $-\alpha^{3.7III}$ deletion arose in northern Melanesia and was picked up by the settlers of Polynesia on their way into the central Pacific; this means that there must have been genetic interchange between the two groups, though this has not yet been quantified.

While the distribution of the $-\alpha^{3.7III}$ deletion gives the best evidence to date that the Polynesian colonists not only passed through Melanesia on their way eastward but spent enough time there to pick up some of the genetic characteristics of the Melanesian inhabitants, the triple ζ arrangement tells us something about the Polynesian ancestors. All the triple ζ alleles in Oceania have the same chromosome structure, suggesting that they have a single ancestor. The distribution of this variant, unlike the $-\alpha^{3.7III}$ deletion, is not restricted to the Pacific; it has also been reported in Southeast Asia (in particular in China and Thailand) (Winichagoon *et al.*, 1982; Trent *et al.*, 1986; Hill *et al.*, 1987; Hertzberg *et al.*, 1988), the putative homeland of the Polynesians. Significantly, the Southeast Asian triple ζ chromosomes have the same haplotype as in Polynesians (O'Shaughnessy *et al.*, 1990).

The triple ζ chromosome is also found throughout Micronesia (O'Shaughnessy *et al.*, 1990), forming a genetic bridge between Southeast Asia and Polynesia and implying that there may have been a second route from the mainland to the eastern Pacific. Movement in the opposite direction (from Polynesia to Southeast Asia) is not only *a priori* unlikely, but is argued against by the finding of a rare polymorphism on about 10% of triple ζs in Polynesia (Hill *et al.*, 1987; O'Shaughnessy *et al.*, 1990). This polymorphism is found only in the Pacific, suggesting that it arose after the arrival of the triple ζs into Polynesia (in the same way as HbJ^{Tongariki} occurred on the earlier $-\alpha^{3.7\text{III}}$ deletion) but after leaving mainland Southeast Asia since the superimposed polymorphism has not been found there.

The relationship of Polynesians to Micronesians and Melanesians can also be inferred from the distribution of the α-globin haplotypes. In most parts of the world haplotypes of classes I and II are common (Table 5.2) (Higgs *et al.*, 1986). Melanesia is almost unique in possessing much higher frequencies of classes III, IV and V. They account for 92% of haplotypes in Papua New Guinea and 88% in island Melanesia. In contrast, they make up less than 10% of haplotypes in Southeast Asia, where 90% are classes I and II (Table 5.3). Polynesia is midway between these two patterns. In eastern Polynesia classes I and II comprise 60% of the total, 40% belonging to classes III to V. Micronesians have a distribution of haplotypes similar to the Polynesians (O'Shaughnessy *et al.*, 1990), though in both regions there is much local variation, some of which is obviously attributable to migration. Thus, in general, Micronesian islands close to Melanesia have higher frequencies of Melanesian haplotypes. For instance, Ponape and the Gilbert Islands have Melanesian components of between 35 and 48% while Palau and Majuro have between 11 and 28% (O'Shaughnessy *et al.*, 1990).

The expectation from the available evidence is that the Polynesians must have experienced numerous population bottlenecks in their colonization of the Pacific; but protein and antigen polymorphism data have not detected the loss of genetic heterozygosity that this process implies. A number of features of the globin data could, however, be attributable to bottlenecks; the total number of α-globin and β-globin haplotypes in eastern Polynesia is less than in western Polynesia, and much less than in Melanesia; the extent of inter-island genetic variation would be compatible with bottlenecks eliminating some variants and increasing the frequency of others. Work on the distribution of alleles at other loci supports this view. The phenylalanine hydroxylase gene haplotypes show increased homozygosity in the most recently colonized Polynesian islands (Hertzberg *et al.*, 1989a) and the finding of a single mitochondrial DNA lineage in 93% of eastern Polynesians is evidence of a bottleneck in at least the female colonists (Hertzberg *et al.*, 1989b).

Table 5.3. *Gene frequencies of α-globin haplotypes in Oceania*

Haplotype	Micronesia	PNG Highlands	Vanuatu	Fiji	West Polynesia	East Polynesia	Brunei	Burma	Thailand	S China
I a	27.7	0	2.6	22.0	27.7	39.3	11.4	31.3	19.4	33.3
c	2.1	0	0	1.7	0	0	0	0	0	1.8
d	0	0	0	1.7	0	0	5.7	0	5.6	5.3
II a	11.5	0	0	0	5.3	6.1	20.0	50.0	30.6	28.1
c	12.0	0	0	0	2.9	1.6	17.1	0	5.6	1.8
d + e	15.2	0	7.9	11.9	26.2	14.1	28.6	9.4	19.4	17.5
f + g	2.6	0	0	3.4	0	0	0	3.1	2.8	1.8
III a + b	13.6	78.3	44.7	22.0	20.1	9.7	14.3	3.1	16.7	7.0
c	3.7	0	0	0	0	0.4	0	3.1	0	3.5
e	0.5	0	0	3.4	1.5	1.6	2.9	0	0	0
h	0	0	0	0	1.5	0	0	0	0	0
IV a + b	7.3	13.0	44.7	16.9	13.6	21.1	0	0	0	0
V b	1.0	0	0	3.4	0.3	0.4	0	0	0	0
c	2.6	8.7	0	13.6	0.3	3.2	0	0	0	0
Number of chromosomes	191	23	38	59	339	247	35	32	36	57

West Polynesia includes the islands of Tonga and Samoa; East Polynesia includes the islands of Niue, Tahiti, The Cook Islands and New Zealand Maoris. Samples from South China include some from Hong Kong. Sources: Flint *et al.*, 1986; Hertzberg *et al.*, 1988; Hill *et al.* 1987; O'Shaughnessy *et al.*, 1990.

One study using six minisatellite loci (unlinked to either globin gene cluster) has now demonstrated that Polynesians have a lower genetic heterozygosity than Melanesians (Flint *et al.*, 1989). This new approach to the analysis of population structure has yet to be fully explored but it is likely that over the next few years our understanding of the structure of Pacific populations will expand dramatically.

Assuming that the globin gene clusters are representative of the human genome, many features of the genetic structure of Polynesian populations can be explained by the likely genetic make-up of their ancestors and the consequences of their migration into Polynesia. The unusual triple ζ variant and the α-globin haplotypes represent an inheritance from their Southeast Asian homeland, while the presence of the $-\alpha^{3.7\text{III}}$ deletion is a consequence of their passage through Melanesia. A number of bottle-necks occurred during the process of colonization and these are attested to by a reduced heterozygosity in Polynesians in comparison to other world populations.

Preliminary work on Australian aborigines provides a further example of the effects of population movements. Genetic studies of red blood cell antigens, enzyme systems and serum protein variants had led to the belief that Papua New Guinea highlanders and Australian aborigines had a common ancestor about 40 000 years ago (Kirk, 1979). Analysis of the α-globin cluster questions such an interpretation; the Australians have a $-\alpha^{3.7\text{II}}$ deletion (which is not found in Papua New Guinea), a $-\alpha^{3.7\text{I}}$ deletion in association with a haplotype different from that found in Papua New Guinea and, most unusually, a $\zeta\zeta\zeta$ chromosome that has a unique haplotype, unlike that found anywhere else in the Pacific (Yen-chitsomanus *et al.*, 1986a,b; Tsintof *et al.*, 1990). The explanation for these findings is not yet clear; it is very likely that there has been a more complex colonization of Australia than initially suspected.

This description of the distribution of globin variants in the Pacific in terms of genetic drift and population movements is lacking in one respect: it has not explained why some genes reach remarkably high frequencies in some areas but not others. Genetic drift and founder effects cannot easily explain the contrast between an α-thalassaemia gene frequency of 70% on the north coast of Papua New Guinea and one of 4% in an otherwise genetically similar population only a few hundred miles away in the high interior of the island (Flint *et al.*, 1986). Fortunately molecular genetic analysis does not fail us here; it can provide good evidence that selection is responsible for the high frequencies of α-thalassaemia, and indeed of other haemoglobinopathies elsewhere in the world.

Malaria: mid-19th century

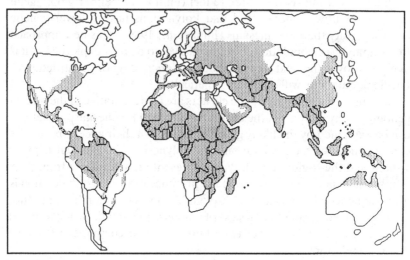

Fig. 5.5. The geographical distribution of malaria before the eradication campaigns of this century (O'Shaughnessy *et al.*, 1989).

Genetic diversity: selection

Despite the undoubted importance of selection in determining the genetic structure of populations, it has proved extremely difficult to demonstrate that it has operated on humankind. Before the introduction of DNA technology the best evidence came from work on sickle-cell anaemia and β-thalassaemia. Both these diseases occur in areas where malaria is or has been a major cause of death. The epidemiological association between haemoglobinopathies and *Plasmodium* infection (Figs 5.5, 5.6 and 5.7) was the first piece of evidence in favour of the malaria hypothesis: that is, the idea that the genetic disorders of haemoglobin synthesis in some way protect their bearers from the worst consequences of infection by the malaria parasites (Haldane, 1948). In the case of the sickle-cell gene, the argument was advanced that since homozygotes inevitably died young, in the vast majority of cases before reproductive age, the observed frequency of haemoglobin S (HbS, the protein product of the sickle cell gene) could be explained only by a remarkably high mutation rate or by the selective advantage accruing to the heterozygotes. In the latter case, selective advantage balances the gene loss caused by the early death of homozygotes. Evidence against the existence of a high mutation rate for HbS was relatively easy to find; the major problem was deciding which of the unfortunately numerous human

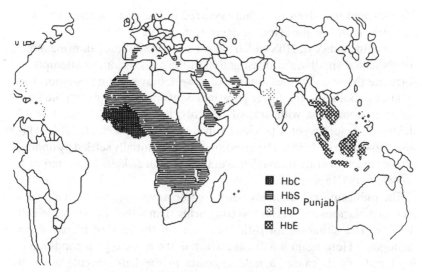

Fig. 5.6. The geographical distribution of the common abnormal haemoglobins: HbS, HbC, HbF, and HbD^{Punjab} (O'Shaughnessy *et al.*, 1989).

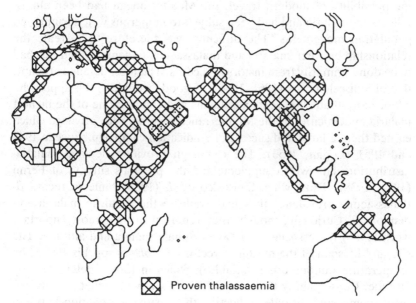

Fig. 5.7. The geographical distribution of the thalassaemias (O'Shaughnessy *et al.*, 1989).

illnesses and misadventures that occurred in the areas where HbS was prevalent could be the selective agent responsible.

The coincidence of HbS and malaria was an initial clue, allowing more sophisticated small-scale investigation of the association, in attempts to correlate the two. Various kinds of data were drawn upon to support the malaria hypothesis: the lower parasite counts in HbS heterozygote trait children, compared with normal controls, and an age stratification of sickle-cell heterozygotes (reviewed in Rucknagel and Neel, 1961; Allison, 1964; Jonxis, 1965). The question was substantially settled by finding a lower mortality rate from falciparum malaria for sicklers compared with non-sicklers (Raper, 1956).

The published evidence for heterozygote advantage of the thalassaemias in malarious areas is far less convincing than is the case for sickle-cell disease. The situation for β-thalassaemia on the face of it may seem analogous. Here again is a disease which in the homozygous condition is frequently fatal, causes a mild anaemia in the heterozygote and yet reaches gene frequencies of 15% in many malarious areas of the world (Livingstone, 1985). The difference with HbS is that, initially at least, reports of β-thalassaemia came predominantly from the Mediterranean. Unlike Africa, where malaria was still uncontrolled at the time researchers arrived and the indigenous population had not availed itself of the possibility of modern travel, the Mediterranean had been almost relieved of malaria and had been subject to numerous well documented population movements. The difficulty was therefore to recreate the relationship between malaria and thalassaemia endemicity and this had to be done primarily from historical records. The most detailed attempt, that of Siniscalco et al. (1961, 1966) is a good example of this approach.

For more than 2000 years Sardinians had endured one of the highest malaria endemicities in the Mediterranean, an onerous privilege which earned them a large and successful eradication campaign between 1947 and 1951 (Logan, 1953). The campaign provided detailed malarial distribution data which supplemented the previous surveys of Fermi (1934, 1938). Thus, when Siniscalco et al. (1961) came to record β-thalassaemia prevalence they had available the requisite malaria epidemiology. Curiously, though, they reported in their most important work (1966) the association between β-thalassaemia and altitude. The size and lifespan of the mosquito vector (Anopheles spp.) is altered by temperature (among other variables), which in turn is determined by altitude. Consequently there should be a negative correlation between thalassaemia and altitude, reflecting the positive correlation between thalassaemia and malaria. This is exactly what Siniscalco *et al.* (1966) claim to have found. However, their study, often quoted as the best evidence for the malaria hypothesis as applied to β-thalassaemia, has not gone unquestioned. Indeed, one author (Brown, 1981) shows that malarial prevalence, as presented in the surveys of the 1930s, does not correlate, at least in a simple way, with altitude. Some of the highest malaria rates are found in the inland high plateaux, while the distribution in the lower plains is mixed. Moreover, as had been the case with HbS investigations, some studies of β-thalassaemia in other areas of the Mediterranean failed to substantiate the malaria hypothesis (Stama-toyannopoulos and Fessas, 1964; Plato *et al.*, 1964).

What can the new technology add to our understanding of the relation-ship between haemoglobinopathies and selection? A major contribution has been the demonstration that a haemoglobinopathy virtually ignored in earlier studies of the malaria hypothesis achieves high frequencies because of selection by malaria. The haemoglobinopathy is α^+-thalassaemia; the reason its epidemiology had not been previously investigated was the inability of standard haematological techniques to diagnose the condition accurately. But it has not been merely the diagnostic capacity of DNA analysis that has provided evidence of malarial selection. Just as important has been the ability to characterize mutations by haplotype analysis, telling us about their origins in a way that has already been exemplified above in the studies on Pacific islanders.

Malaria and α^+-thalassaemia

There are a number of problems in investigating the relationship between α^+-thalassaemia and malaria that make it an even more difficult under-taking than is the case for HbS and β-thalassaemia. One has already been mentioned: it needs DNA analysis to determine the genotype correctly, so that measures of α^+-thalassaemia prevalence have only recently become possible. A further difficulty arises because, since even homo-zygous α^+-thalassaemia is a mild condition, the selective agent respon-sible for the increase in gene frequency need not be severe; therefore a large number of factors could be responsible, far more than had to be considered in investigations into HbS and β-thalassaemia.

A third complication is that α^+-thalassaemia may ameliorate the effect of other haemoglobinopathies, as has been suggested to occur in its interaction with the sickle cell gene (Higgs *et al.*, 1982; Pagnier *et al.*,

1984b; Embury 1985; Kulozik *et al.*, 1988). Thus the epidemiological demonstration that malaria has selected the α^+-thalassaemia gene requires a survey by gene analysis of an area with well documented malaria prevalence rates, the population of which is free of other haemoglobino-pathies.

One area which satisfies some of these conditions is the Southwest Pacific. Although control programmes have altered malarial endemicities, numerous surveys were carried out prior to intervention (Lambert, 1949; Black, 1954). These show that malaria prevalence varies with both altitude and latitude; it is absent above 2000 metres in Papua New Guinea, and also beyond 170° E. The area is also relatively devoid of other haemoglobinopathies (though not of globin variants). Neither HbS nor HbE has been found and, apart from a few islands, β-thalassaemia is rare (Livingstone, 1985; Bowden *et al.*, 1985). Analysis of more than 1800 DNA samples showed that the frequency of α^+-thalassaemia correlated with malaria in both altitudinal and longitudinal clines, while a variety of other unlinked DNA polymorphisms did not (implying that the process responsible for the α^+-thalassaemia distribution is locus specific) (Flint *et al.*, 1986).

Haplotype analysis in Melanesia allows the origin and fate of mutations to be deduced and this greatly adds to the argument in favour of malarial selection. In Papua New Guinea, the commonest deletion is the $-\alpha^{4.2}$ type, and the great majority of these occur on a single haplotype $(--+MZ-)$ (Flint *et al.*, 1986). This is also the commonest $\alpha\alpha$ haplotype in Melanesia and happens to be very rare in other world populations. In particular, it is uncommon in neighbouring Southeast Asian populations where the $-\alpha^{4.2}$ deletion also occurs but is found on a different haplotype. This means that the α^+-thalassaemia in coastal Papua New Guinea probably arose locally (and therefore cannot have been imported from Southeast Asia) and a local mechanism must have elevated it to its present very high frequencies. A similar argument can be made for α^+-thalassaemia in Vanuatu. Here it is mostly of the $-\alpha^{3.7III}$ variety. As mentioned earlier, this type is itself unique, not having been detected elsewhere in the world (apart from Polynesia), and haplotype analysis strengthens the case for its local origin. It too arose on a $--+MZ-$ haplotype (Hill *et al.*, 1985a); but unlike the $-\alpha^{4.2}$ deletion it is found only in association with this haplotype, implying that is has arisen only once.

Altogether four different deletions are responsible for α^+-thalassaemia in Melanesia and, counting the same deletions that occur on different haplotypes as independent mutations, six α-globin gene deletions have risen to relatively high frequencies in Melanesia (Flint *et al.*, 1986). Although small differences in haematological indices have been noted between the $-\alpha^{3.7III}$ and the $-\alpha^{4.2}$ types (Bowden *et al.*, 1987), none has been noted between the other deletions, and none is known for the same deletion occurring on different haplotypes. It appears that six mutations, producing practically identical phenotypes, have all been maintained in the Melanesian population.

There are therefore a number of epidemiological reasons why selection must be responsible for the high frequencies of α^{+}-thalassaemia in Melanesia: the genes are distributed in latitudinal and altitudinal clines (not found for other unlinked DNA polymorphisms), a distribution which mutation and drift alone cannot readily account for; multiple genotypes are associated with an almost identical phenotype, implying that the process altering the gene frequency acts on the phenotype; these genotypes arose locally, and different mutations predominate in different Melanesian populations that are otherwise genetically almost indistinguishable, making alternative explanations for their presence based on population movements virtually impossible to maintain. The correlation with malaria in both altitude and latitude indicates that this disease is the agent responsible, but has not indubitably established the fact. We still lack good *in vitro* evidence of the protective effect of α^{+}-thalassaemia and much could still be learnt from studies of malaria mortality rate of those with and without α^{+}-thalassaemia.

The study of α^{+}-thalassaemia provides a picture of the epidemiology of a gene under selective pressure in a human population; multiple mutations have produced an almost identical phenotype, and these mutations are regionally specific. Do the other haemoglobinopathies display similar epidemiologies? The most exhaustive investigation of a mutant's origin has been carried out for the sickle-cell gene; here the question of malarial selection has been substantially settled, but only molecular studies have been able to define the number of mutations and their origins.

Origins of the sickle cell gene

The protein of HbS is always the same whether it is found in India or Africa, but does that mean it is the same gene, that it has originated in one area and spread to others, or has it arisen more than once in different world populations? Molecular genetic analysis of the population distribution of HbS was first attempted with one of the first DNA polymorphisms discovered, the Hpa I site 3' to the β gene. Kan and Dozy (1978) found that HbS occurred in association with both Hpa I alleles, and argues that the present African and Saudi sickle cell anaemia had a dual origin; however, Solomon and Bodmer (1979) argued from the same data that there was a single, but ancient, origin for HbS, with subsequent recombination that shifted the mutant to a second chromosome. To decide whether the haplotype diversity was a product of recombination or reflected multiple origins, a more detailed haplotype was required. If Solomon and Bodmer's view was correct, then the number of haplotypes found bearing HbS should increase with the number of new β-globin polymorphisms discovered. A study using seven polymorphic sites found that HbS in Jamaicans occurred in association with no fewer than 15 different haplotypes (Table 5.4) (Wainscoat et al., 1983b). Did this incontrovertibly mean that there had been a single origin?

In fact the authors argued that the HbS probably had multiple origins. They drew this conclusion from the observed linkage disequilibrium between 5′ and 3′ parts of the β-globin haplotype found for HbS alleles. As discussed above, there is increased recombination between the two sub-haplotypes, and in most world populations there is random association between the two parts; not so in the Jamaican HbS chromosomes, where there is strong linkage disequilibrium between 5′ and 3′ haplotypes. A similar study by Antonarakis et al. (1984a) using 11 polymorphic sites identified 16 haplotypes bearing HbS in American Blacks. They also noticed the 5′–3′ haplotype linkage and pointed out that the haplotypes could be divided into four groups on the assumption that haplotypes were derived from each other by a single recombination. They found HbS on three different β-globin gene sequences, which would be hard to explain as simply the result of recombination with a single mutant. Although this made a unicentric hypothesis difficult to maintain, it could not resolve how many mutations had occurred.

One way to decide was more detailed examination in different areas of Africa. Since local migration is still limited and the chance of a mutation occurring more than once in the same locality is very low, one prediction was that a single, or at best a handful, of HbS haplotypes would occur in an area, and that in neighbouring areas a different haplotype would be prevalent. Pagnier et al. (1984a) analysed 124 African HbS chromosomes and found six different haplotypes. When they assigned these haplotypes geographically, a remarkable pattern of genetic homogeneity emerged. In each of the four regions they examined (Senegal, the Central African Republic (CAR), Benin and Algeria), one haplotype was predominant; indeed in Benin and Algeria, HbS occurred on only one haplotype (Table

5.4). This regional homogeneity and the fact that differences between the three common haplotypes were found at both the 5' and 3' ends led the authors to postulate that the African HbS mutation had arisen three times. A more detailed study of the DNA sequence immediately 5' to the β gene by the same group gave further support to the tricentric thesis (Chebloune *et al.*, 1988). The sequence here is characterized by a number of ATTTT and $(AT)_xT_y$ repeats (Spritz, 1981; Semenza *et al.*, 1984). The Benin, Senegal and CAR HbS haplotypes had repeats that were roughly homogeneous and again regionally specific, thus making it appear unlikely that a single mutant could have spread to the three haplotypes through recombination.

This leaves unresolved the relationship of African HbS to that found elsewhere in the world. Haplotype analysis suggests that the Mediterranean HbS was derived from Africa; not only is the Algerian and Benin haplotype identical, but this haplotype is found on all Mediterranean sickle-cell chromosomes. What, however, is to be made of the finding that HbS occurs on three different β gene sequences in American Blacks? Again, more specific analysis casts light; Indians and Saudis share a HbS haplotype (Kulozik *et al.*, 1986); this is different from the African haplotypes of Pagnier *et al.* (1984a) and occurs on a different β gene sequence, but is the same as one reported to occur in American Blacks (Antonarakis *et al.*, 1984a).

It appears from this evidence that four mutations have produced HbS, but one drawback to the multicentric origin theory remains: why should three mutations occur in Africa and none in Southeast Asia, which has just as much endemic malaria? One answer is that the interaction with pre-existing haemoglobinopathies may be detrimental to the gene's chances of becoming widespread in the population. Comparing the distribution of HbS with other haemoglobinopathies is illuminating here. The relative absence of β-thalassaemia from areas of Africa with high HbS prevalence has been explained by the poor fitness of the HbS/β-thalassaemia compound heterozygote in comparison to the heterozygote of either mutation (Livingstone, 1985), an example therefore of the interaction theory mentioned above. However, β-thalassaemia is caused by at least 100 different mutations (Kazazian and Boehm, 1988), almost none of which have separate origins in the way proposed for HbS. Why then should codon 6 of the β chain be so favoured as to receive the same mutation three times and produce HbS in three neighbouring areas of Africa? The distribution of haemoglobins E and C provides an interesting comparison and suggests an answer.

Table 5.4. *Percentage gene frequencies of β-globin haplotypes on which the sickle cell gene is found*

Haplotype	US Black %	Jamaica %	Nigeria %	Benin %	Algeria %	South Africa %	CAR %	Senegal %	Saudi Arabia %	India %	Portugal %	Total %
+ - - - + +							3.4	1.8	9.7	2.2		1.6
- + -	1.2											0.2
+ - +												0.0
- + + - + +			2.9							2.2		0.3
- + -												0.0
+ - + + +		1.4					3.4					0.7
- + - + +	7.1	4.6					3.4	80.7			25.0	10.9
- + -										4.4		0.3
+ - + + +		0.9							1.4			0.5
- + + + +		1.4						1.8				0.7
+ - +												0.0
- + + +												0.0
- + - - + + +	74.1	70.8	97.1	100.0	100.0		6.9	14.0	23.6		12.5	52.1
+ - +												0.0
- + +		0.5										0.2
- - + + +												0.0
- + - + +		0.5										0.2
+ - +												0.0
- + +												0.0
- + -												0.0
+ -		0.5										0.2

Haplotype	n=85	n=219	n=20	n=20	n=34	n=57	n=72	n=20	n=29	n=45	n=16	Total (n=614)
- + - - - + +	17.6	15.5										16.6
- + -												0.0
+ - -												0.0
+ - - - + +		1.4										0.5
- + -												0.0
+ - + - + +												0.2
- + -		0.5										0.5
+ - -												0.0
+ + - - - + +						82.8	62.5		100.0	86.7	43.8	13.8
- - - + + + -		0.9										0.3
+ - - - - - -		0.5				1.8	2.8			2.2		0.2
												0.2
		0.5										0.5
Unclassified										2.2	18.8	
Totals	85	219	20	20	34	57	72	20	29	45	16	614

CAR, Central African Republic. Sources: Antonarakis et al., 1984a; Chebloune et al., 1988; Kulozik et al., 1986; Monteiro et al., 1989; Pagnier et al., 1984a; Ramsay and Jenkins 1987; Wainscoat et al., 1983b.

Origins of haemoglobins E and C

Analysis of the origin of another common globin gene variant, HbE, provides a striking parallel with HbS. HbE is the second commonest haemoglobin variant known, and is found throughout Southeast Asia (Livingstone, 1985), again in association with malaria. It produces a mild anaemia in the homozygote and reaches high frequencies in some areas of endemic malaria. Antonarakis *et al.* (1982a) found three haplotypes among 23 HbE chromosomes and among these only two different 3′ haplotypes ('+ −' and '− +'). More recent work on larger numbers of HbE chromosomes confirms that there are at least two 3′ haplotypes (Nakatsuji *et al.*, 1986; Yenchitsomanus *et al.*, 1988; Hundrieser *et al.*, 1988; Yongvanit *et al.*, 1989). The conclusion reached from these data is that the situation is analogous to HbS, and that HbE might have at least two origins. Also like HbS, HbE mutations occur in neighbouring geographical regions; HbE is absent from the Mediterranean and Africa (though there are cases of HbE occurring apparently as a new mutation in Europe: Kazazian *et al.*, 1984c). The epidemiology of HbE therefore poses the same question as that of HbS: why should multiple mutations producing the same genotype occur in relatively circumscribed areas? Apart from the interaction theory, there still remains the possibility that the mutation has arisen once and spread by gene conversion onto new haplotypes. Gene conversion is a process whereby one sequence is homogenized by another; it results in the non-reciprocal transfer of sequence information (unlike recombination, where there is reciprocal exchange). The conversions needed to spread the HbS mutation from one chromosome to another must have been very small, because in order for the codon 6 mutation to have been passed onto a new haplotype without also transferring the nearby $(AT)_xT_y$ repeats, less than 2 kb of DNA must have been converted. Nevertheless, small conversions are known to occur (Smithies and Powers, 1986; Starck *et al.*, 1990; Borts and Haber, 1989). Fourteen gene conversions have been postulated to have occurred in the recent evolution of primate γ-globin genes (Slightom *et al.*, 1988), and the frequency of these conversions is thought to be higher than that of recombination across the β-globin gene cluster (Smithies and Powers, 1986; Starck *et al.*, 1990). Since the HbS and HbE alleles have reached high frequencies in their respective localities, it is therefore possible that small gene conversions may be responsible for the presence of these mutations on more than one haplotype.

The origin of haemoglobin C (HbC) has also been examined by haplotype analysis. Like HbE it produces a mild anaemia in the homozygote. It is found almost exclusively in malarious West Africa (Livingstone, 1985). Boehm *et al.* (1985) found that 22 of 25 HbC alleles had the same 5' and 3' haplotypes; all mutants had identical 3' haplotypes as would be predicted by a theory of single origin with recombination producing spread to new haplotypes. The occurrence of HbC in a relatively small area and the fact that linkage is still found in almost 90% of cases with the 5' haplotype suggests that the mutant is recent in comparison to HbS and HbE. Perhaps, given time, conversions will occur that will distribute the mutation onto different 3' haplotypes to give an epidemiology similar to HbS and HbE.

Malaria and β-thalassaemia

Malarial selection of β-thalassaemia has not been examined so thoroughly with molecular techniques as has α-thalassaemia, but a powerful argument in favour of the selection hypothesis can be drawn from the large amount of data collected in the course of feasibility studies for the antenatal diagnosis of β-thalassaemia. Such studies aim to determine how many types of β-thalassaemia are prevalent in areas where the β-thalassaemia gene frequency is high. If there is a limited number of mutations that can easily be detected, then antenatal diagnostic facilities are relatively easy to establish. Consequently the investigations provide a remarkably detailed picture of the distribution of β-thalassaemia worldwide, and reveal much unexpected information about population structure.

Two important conclusions can be drawn from these studies: (1) that there is a close, but not invariant, association between particular β-thalassaemia mutations and haplotypes, and (2) that mutations are regionally specific. These findings can be best explained by the action of selection.

The first investigators of the molecular defects of β-thalassaemia concentrated their efforts on the Mediterranean. Orkin *et al.* (1982) began a systematic sequencing study of cloned β-thalassaemia genes. The first three genes they sequenced had identical mutations; they also had identical haplotypes. Their work, on a relatively small number of genes (26) seemed to support the view that different β-thalassaemias occur on different haplotypes; with only a couple of exceptions, mutations were linked to haplotypes. But, if each haplotype carried only one mutation, since only three haplotypes constituted more than 90% of the total number of haplotypes in the population, could it be true that only a small number of mutations linked to these common haplotypes were respon-

sible for the vast majority of β-thalassaemia? Quantifying the association
between haplotypes and mutations became imperative, but at that time
only the small number of mutants that disrupted a restriction enzyme site
could be easily detected; the remainder required time-consuming sequencing for detection. Fortunately the advent of oligonucleotide analysis made it possible to determine the mutation associated with each
haplotype (Conner et al., 1983). Once the mutant sequence is known, an
oligonucleotide can be manufactured that anneals specifically to the
variant part of the β-thalassaemia gene (but the technique cannot
determine if other mutations are present). When the oligonucleotide fails
to detect its complementary sequence, and traditional haematological
investigations show that the gene produces β-thalassaemia, cloning and
sequencing can detect the new mutation. A further oligonucleotide is
then made to determine the frequency of the newly discovered mutant.
By continuing this process all mutations should eventually be characterized in a population and their frequencies become known.

What does oligonucleotide analysis of β-thalassaemia frequencies
reveal? Looking in more detail at their Mediterranean patients, Kazazian
et al. (1984b) determined the mutation in 156 of 162 β-thalassaemia
genes. Nine mutations were found of which two (the IVS1 110 and
nonsense 39 mutations) accounted for 34% and 28%, respectively, of the
total. The frequency distribution of mutations in fact mirrored that of
haplotypes: a small number form the great majority, and a larger number
are present at lower frequencies. As might be expected, the common
haplotypes carried more than one mutation; for instance, the most
common 5' haplotype (+ − − − −) occurs in association with seven
different mutations (Table 5.5).

However, if a more extended haplotype of seven sites is used, including
two sites of the 3' haplotype, then each mutation is found in association
with fewer haplotypes. This is because there is almost invariant association between 3' haplotypes and β-thalassaemia mutations; for instance,
the nonsense codon 39 β-thalassaemia variant occurred on 3' haplotype
+ +; this haplotype is found in Mediterraneans in association with three
5' haplotypes (− + − + +, − + + − + and + − − − −); the nonsense 39
mutant is accordingly found on all three , the commonest being + − − − −
(i.e., the full seven-site haplotype on which it occurs is + − − − − + +). In
only two cases were the same mutations found on different 3' haplotypes
(Kazazian et al., 1984b). Similar studies carried out elsewhere in
the world showed the same picture, although more cases are now
documented of the same mutation occurring on different 3' haplotypes
(Table 5.5).

The second important finding from the surveys of β-thalassaemia incidence is the regional specificity of mutations. Although the sample is still weighted in favour of Europeans, the spectrum of β-thalassaemia mutations found in different areas of the world is now fairly well known. Indeed, some authors have even stated that virtually all the mutations that produce β-thalassaemia are known (Wong *et al.*, 1986). Table 5.6 displays the haplotypes of normal and thalassaemia chromosomes for different world populations so far studied. The skewed distribution found for normal β-globin chromosome is repeated in the distribution of β-thalassaemia haplotypes, but there are differences. In Europe, 33% of thalassaemia chromosomes are $-++-+++$, but this haplotype is found on only 9% of normal β-globin chromosomes; in India, 49% of thalassaemia haplotypes are $+----\ --$ in contrast to 12% of normals. These differences are significant at the 0.001% level. This may be partly attributable to the biases of the β-thalassaemia studies. More than one group has reported results from Sardinia where the nonsense 39 mutation is common and occurs on the $-++-+++$ haplotype, the frequency of which is consequently elevated. However, it is also important to note that the differences are regionally specific. The reason for this is that the haplotypes bearing β-thalassaemia mutations are not the same in different parts of the world.

Much more detail is provided by the distribution of the individual mutations (Table 5.5) where the regional specificity becomes apparent. In Europe the IVS1 110 mutation is common, but unknown in China. Conversely, in China the codon 41 TCCT deletion is common but unknown in Europe. Both mutations occur on the same haplotype $(+----\ ++)$. The difference in the spectrum of mutations is remarkable; when the first thalassaemia genes were sequenced from India, seven new mutations responsible for β-thalassaemia were found (Kazazian *et al.*, 1984a).

To explain these findings it can be argued that the majority of mutations have arisen only once; their presence on multiple 5' haplotypes is because of recombination between 5' and 3' haplotypes. The most extreme example is to be found in Sardinia, where a total of five different seven-site haplotypes bear the nonsense 39 mutation but all share the same 3' haplotype $(++)$ (Pirastu *et al.*, 1987). Since the latter has a frequency of about 50% in the population of Sardinia the mutation may well have occurred more than once on the same haplotype. Pirastu *et al.* (1987) explored this possibility by using six polymorphic sites in the vicinity of the β-globin gene. They showed that, with the exception of two chromosomes, the haplotype was the same $(++++-)$. Even the two

Table 5.5a. *Association between β-globin haplotypes and β-thalassaemia mutations in different world populations. For sources see Table 5.1*

Haplotype	Fr 9 Turkey	IVS1 110 North Italy	South Italy	Cyprus	Sardinia	Algeria	Tunisia	Lebanon	Spain	Portugal	Turkey
+ - - - + + / - + + / +	100	100	90.4	100	100	100	100	100	100	100	93.2
- + + + + / - + +			3.8								4.1
+ - + + / - + + / +			5.8								1.4
- + + / + - + +											1.4
+ + - - + / + - - + -											
+ + + - + / + + + - -											
- + - + - / - + - - - +											
Totals	1	32	52	32	2	17	5	30	3	3	73

Table 5.5a. (*cont.*)

Haplotype	Nonsense 39									IVS 1:1					
	North Italy	South Italy	Sardinia	Algeria	Tunisia	Lebanon	Spain	Portugal	Turkey	South Italy	Algeria	Tunisia	Spain	Portugal	Turkey
+ − − − + +	9.3	30.5	19.6	33.3	14.3		34.3	26.7		100	50.0	100.0	100.0	100.0	100.0
− +		1.7													
+ −	1.9														
− + − + + +	51.9	52.5	66.7	55.6	42.9	100	60.0	66.7	50.0		50.0				
− +							5.7								
+ −			11.1												
− + − + + +	35.2	15.3	9.1						50.0	50.0	50.0				
− +															
+ −			1.8												
− + − − + +	1.9		2.7		14.3			6.7							
− +															
+ −															
− + + − − + +															
− +															
+ −															
− + − + − +															
− − − + +					7.1										
+ + + − + +					7.1										
− + − − + + +					14.3										
− + −															
− +															
Totals	54	59	219	9	14	2	35	15	4	12	4	1	2	9	4

Table 5.5a. (*cont.*)

Haplotype	IVS 1:6						IVS 2:745				
	South Italy	Algeria	Tunisia	Lebanon	Spain	Turkey	South Italy	Sardinia	Tunisia	Lebanon	Turkey
+ − − − − + +	21.7			20.0		8.3	100.0	100.0	100.0	100.0	100.0
− + + − + +				20.0	28.6	16.7					
− + − + + +		100.0									
− + − − + +	78.3		100.0	60.0	71.4	75.0					
Totals	46	2	7	5	7	24	10	2	5	2	6

Table 5.5a. (*cont.*)

Haplotype	Fr 6						Fr 8		IVS 1:5		IVS 2:1		
	South Italy	Sardinia	Algeria	Tunisia	Spain	Turkey	Lebanon	Turkey	Lebanon	Turkey	Lebanon	South Italy	Turkey
+ − − − + +	100.0	9.1	7.7		33.3								28.6
− +													
+ −				18.2	66.7	100.0		5.6					
− + + − + +		90.9	84.6	72.7			50.0	94.4				100.0	
− +													
+ −													
− + − + + +							50.0		100.0		100.0		71.4
− +										100.0			
+ −													
− + + − + +													
− +													
+ −													
− + − + + +			7.7	9.1									
− +													
+ + + − − +													
− − − + − +													
− +													
− + − − − +													
− − +													
Totals	1	11	13	11	3	2	4	18	2	2	2	3	7

Table 5.5a. (*cont.*)

Haplotype	-87 South Italy	-87 Turkey	29C/T Lebanon	Unknown North Italy	Unknown South Italy	Cyprus	Sardinia	Algeria	Tunisia	Turkey
+ - - - - + +	100.0	50.0			55.6	25.0		31.8	42.3	62.5
- +					11.1			36.4		25.0
+ -			100.0	66.7	11.1		100.0	4.5	7.7	
- + + - + + +				33.3	22.2				38.5	
- +								27.3		
+ -		50.0				75.0				
- + - + +									3.8	
- +										
+ - +										
- +										
+ - - + +										
- +										
+ -										
+ + + - - + -										
- - + +										
+ + + + - - +										
- + - - - +									3.8	
- +									3.8	12.5
- +										
Totals	1	4	4	3	9	4	1	22	26	8

Table 5.5a. (cont.)

Summary

	Fr 9	IVS 100	Non 39	IVS 1:1	IVS 1:6	IVS 2:745	Fr 6	Fr 8	IVS 1:5	IVS 2:1	−87	29 C/T	Unknown	Totals
North Italy		32	54										3	89
South Italy		52	59	12	46	10	1			3	1		9	193
Cyprus		32											4	36
Sardinia		2	219			2	11						1	235
Algeria		17	9	4	2		13						22	67
Tunisia		5	14	1	7	5	11						26	69
Lebanon		30	2		5	2		4	2	2		4		51
Spain		3	35	2	7		3							50
Portugal		3	15	9										27
Turkey	1	73	4	4	24	6	2	18	2	7	4		8	153
Totals	1	249	411	32	91	25	41	22	4	12	5	4	73	970

Table 5.5b. *Association between β-globin haplotypes and β-thalassaemia mutations in Southeast Asia*

Haplotype	41-TCTT						IVS 2:654			
	South China	China	Thailand	UK Asian	Indian Asian	Indonesia	South China	China	Thailand	Indonesia
+ – – – + +	58.1	75.0	51.0	18.2		100.0	89.5	33.3	80.0	71.4
– +	41.1		37.3				10.5	33.3		28.6
+ –				72.7	66.7					
– + + – + +			2.0							
– + +			2.0	9.1	33.3			33.3		
+ –			3.9							
– + +		12.5	2.0							
+ – + – + +	0.8								20.0	
+ – – + +		12.5	2.0							
Totals	124	8	51	11	3	1	57	3	5	7

Table 5.5b. (*cont.*)

Haplotype	Fr 17				Fr 71			IVS 1:5					
	South China	China	Thailand	Indonesia	South China	China	Thailand	South China	China	UK Asian	Indian Asian	Indonesia	Thailand
+ − − − − + +	93.9	50.0	100.0	100.0	97.7	25.0	25.0	40.0		4.3	86.7	34.4	100.0
− − +						75.0	75.0			52.2		59.4	
+ − −								20.0		34.8	6.7		
− + − + + +	3.0							40.0					
− + −					2.3							3.1	
+ −	3.0												
− + − + + +									50.0				
− + −									50.0				
+ −											6.7		
− + +													
− + −		50.0											
− − − − + +													
+ − + − + +													
+ − + − + +													
+ − + − + +													
− − + − + +													
+ − + − + + −										4.3			
+ + − − + +										4.3			
+ − + + − +													
+ − − + + − +													
− + − − + +												3.1	
Totals	33	2	49	1	44	4	4	5	2	23	15	32	1

Table 5.5b. (cont.)

Haplotype	-28 A/G South China	-28 A/G China	-28 A/G Thailand	-29 South China	619 del UK Asian	619 del Indian Asian	Non 43 South China	Fr 8 UK Asian	Fr 8 Indian Asian	-88 UK Asian
+ – – – – + +										
– +	8.5				100.0	100.0	75.0	70.0	100.0	100.0
+ –										
– + + – + + +	72.3	100.0	100.0	100.0			25.0	25.0		
– +										
+ –										
– + – + + +										
– +										
+ –										
– + +										
– +										
– + – + – + +	19.1							5.0		
+ – + – + – +										
+ – + – + – +										
– + + – + +										
+ – + – + +										
+ + – + – + +										
+ – – + + +										
+ – – + + +										
+ – – – + +										
Totals	47	1	1	4	21	13	4	20	1	2

Table 5.5b. (*cont.*)

Haplotype	Non 15			Fr 16		Cap + 1	IVS 1:1		Indonesia	IVS 1 del
	UK Asian	Indian Asian	Indonesia	UK Asian	Indian Asian	UK Asian	UK Asian	Indian Asian	Indonesia	Indian Asian
+ - - - - + +	40.0		50.0							
- +			50.0							
+ -										
- + - + + +		100.0		100.0	100.0	100.0				
- + -										
+ -										
- + - + + +	60.0						100.0	100.0	100.0	100.0
- +										
+ -										
- + - + +										
- +										
- - - - + +										
- +										
- + - + - +										
+ - - + - +										
+ - + - + +										
+ - + + - +										
- + - + + +										
+ + - + + -										
+ + - + + +										
+ - - + - +										
+ - + + - +										
+ - - + +										
Totals	5	1	4	1	1	2	14	1	6	1

Table 5.5b. (*cont.*)

Haplotype	Codon 30 Indonesia	Codon 35 Indonesia	Unknown South China	UK Asian	Indian Asian	Indonesia	Thailand
+ – – – – ++	100.0	100.0	33.3		25.0	75.0	25.0
– +			41.7		25.0		62.5
+ –					25.0		
– + + – + ++			8.3				
– + – ++ ++			8.3		25.0	25.0	12.5
+ –				50.0			
– – – + ++			8.3				
– – – – + ++				50.0			
– + – + – ++							
+ – + – + ++							
+ – + – + – +							
– + – + – + ++							
+ – + + – + ++							
– – + + – ++							
+ – + – + ++							
+ + – + + +							
+ – + + ++							
+ + – + – + ++							
+ – + + – + ++							
+ – – – – ++							
Totals	1	1	12	2	8	4	8

Table 5.5b. (*cont.*)

Summary

	41-TCTT	IVS 2:654	Fr 17	Fr 71	IVS 1:5	-28 A/G	-29	619 del	Non 43	Fr 8	-88	Non 15	Fr 16	Cap +1	IVS 1:1	IVS 1 del	Codon 35	Un-known	Totals
South China	124	57	33	44	5	47	4		4									12	330
China	8	3	2	4	2	1													20
Thailand	51	5	49	4	1	1												8	119
UK Asian	11				23			21		20	2	5	1	2	14			2	101
Indian Asian	3				15			13		1		1	1		1	1		8	44
Indonesia	1	7	1		32							4			6		1	4	69
Totals	198	72	85	52	78	49	4	34	4	21	2	10	2	2	21	1	1	34	683

Table 5.6. *Seven-site β-globin haplotypes from different world populations. Frequencies are given for both normal (N) and thalassaemic (T) chromosomes. For sources see Table 5.1*

A. Seven-site β-globin haplotypes in Europe

Haplotype	North Italy N	North Italy T	South Italy N	South Italy T	Italy N	Italy T	Sardinia N	Sardinia T	Cyprus N	Cyprus T	Turkey N	Turkey T	Algeria N	Algeria T	Tunisia N	Tunisia T	Lebanon N	Lebanon T	Greece N	Greece T	Spain N	Spain T	Portugal N	Portugal T	Mediterranean N	Mediterranean T	Total N	Total T
+ − − − − + +	26.0	38.6	40.0	35.2		20.1	30.6	18.2	50.0	87.0		44.0	40.6	43.4		23.1		60.8	47.6	45.3		32.0		13.0	37.8	38.9	37.6	37.4
− + −	10.0	2.3	5.8	8.2		1.8	13.2		5.3	2.6		6.0	1.4			7.7		3.9	7.1	5.1					10.8	6.8	8.4	4.5
+ − −	28.0	5.7	9.2	6.9		3.6	14.9		10.5	1.3		7.5	14.5	6.2		4.6			19.0	17.5		8.0		39.1	19.8	13.6	15.8	7.4
− + + − + +	2.0	44.3	16.7	24.8		45.0	5.8	66.7	2.6			6.0	8.7	17.8		12.3		13.7	11.9	13.9		42.0		43.5	7.2	20.4	8.7	26.7
− − + −	2.0	1.1	2.5	0.2			11.6							0.8								6.0					3.3	0.3
+ −			0.8				4.1																				1.1	0.1
− + − + + +	20.0	4.5	12.5	7.9		5.9	9.1	6.1	26.3	1.3		2.2	13.0	18.6		27.7			2.4	3.6					13.5	3.7	12.7	6.5
− − +	2.0			1.5			2.5					12.7	1.4	4.7											3.6	0.6	1.8	2.3
+ −	8.0		10.0	1.5		1.2	1.7					3.7	8.7												7.2	4.9	5.8	1.5
− + − − + + +							1.7	3.8											4.8	2.9							0.7	0.7
− −																		3.9										0.1
+ −																		3.9									0.4	
− + + − + +	1.1		0.2	12.4	0.6	20.7	0.8	5.3	5.3	3.9		16.4	2.9	2.3	3.1	10.8		5.9	7.1	10.2		10.0		4.3		9.3	1.6	8.8
− −										3.9		3.9						3.9									0.5	0.3
+ −										3.9		3.9																0.1
− + + − − + +							2.5							3.1	1.5													0.1
− +	2.0																											
− − − + + +		2.3											4.3	2.3											4.3	2.3	0.5	0.2
+ −																												

	50	88	120	403	169	121	132	38	77	134	69	129	65	51	42	137	50	23	111	162	550	1620
− − − − − + +		0.8										2.9	3.1	3.1							0.5	0.4
− +												1.4	0.8								0.2	0.1
+ −																						
− + − + − +																						0.4
− +																					1.2	
− + − − + +		0.8		0.2	1.2																0.6	0.2
− +				0.2																		0.1
+ −															2.0							
− − + +										1.5			1.5									
− +																						
+ −																						
− − − + + +		0.8																				
− +																						
+ −																						
+ − − + + +				0.2									1.5									0.1
− +																						
+ −																						
+ + + − + +																						
+ + − − + −																						
+ − − − + +				0.2																		
+ − − − −																						
Totals	50	88	120	403	169	121	132	38	77	134	69	129	65	51	42	137	50	23	111	162	550	1620

Table 5.6b. (cont.)

B. Seven-site β-globin haplotypes in Southeast Asia

Haplotype	South China N	South China T	China N	China T	Thailand N	Thailand T	UK Asian N	UK Asian T	Indian Asian N	Indian Asian T	Indonesia N	Indonesia T	Japan N	Japan T	Cambodia N	Cambodia T	Totals N	Totals T
+ − − − − + +	25.6	43.5	26.7	45.7	17.5	30.8	33.8	25.0	31.3	7.6		38.2	38.9		4.3		24.7	35.8
− + +	32.2	34.3	36.7	33.0	26.8	47.2	10.4	33.5	12.5	65.5		33.8	11.1		42.6		24.7	39.1
+ −	20.7	1.5	23.3	2.5	40.2	1.5	5.2	8.5	10.0	4.2			16.7		25.5		20.9	2.8
− + + − + + +	4.1	0.3	3.3	9.7		3.6		0.6	8.8	1.7							2.8	0.6
− +	2.5	14.8	6.7		2.1	2.1		6.1	3.8	1.7					2.1		1.7	8.8
+ −	1.7	0.9															1.7	0.3
− + − + +	4.1	0.3		0.6	6.2	1.5	20.8	19.5	13.8	15.1		25.0	11.1		8.5		9.1	4.9
− +	3.3	0.9			1.0	1.0	10.4	2.4	12.5	1.7			5.6		6.4		5.7	1.0
+ −	1.7				1.0	2.1		0.6		1.7			5.6		4.3		1.1	0.4
− + − − + + +							10.4	2.4									1.7	0.3
− +						1.0	6.5			0.8					4.3		1.7	0.4
+ −						1.0	2.6	2.4									0.4	
− + + − − + +					1.0													0.2
− +	0.3																	
+ −				2.5	1.5	1.5												0.1
+ +				3.3	1.0	1.0												0.2
− − − + + +									2.5									
− +						1.0												1.0
+ +																		1.2
− + − + − + +					1.0	1.0												
− +																		0.2

	121	324	30	361	97	195	77	164	80	119	68	18	47	470	1163
+ −	0.8		3.3		0.6	0.5								0.2	0.1
+ +	0.8				3.1					1.3	1.5			0.6	0.2
− +														0.4	
+ +														0.2	
+ −	1.7				0.6	1.5			1.3					0.2	0.3
− +														0.4	0.2
+ +		0.3				0.5		0.6							0.2
+ −						1.0		0.6		1.3			11.1	0.4	0.2
+ +		2.8		0.3							1.5				0.8
+ +	0.8			0.6		2.1									0.1
− −				0.3											0.1
− −															0.1
− +															0.3
− −															0.2
− −															0.1
Totals	121	324	30	361	97	195	77	164	80	119	68	18	47	470	1163

exceptions differed only by one and two sites, and could be derived from the consensus by gene conversion or point mutation. (Unfortunately, as they did not report the frequency of the six-site haplotypes in the population, it remains possible that $++++-$ is just as common in the normal population as the two-site 3′ haplotype $++$.) As mentioned above, there is such strong linkage disequilibrium between sites in the haplotype that the addition of extra sites does not guarantee greater specificity. Nevertheless, the likely explanation is that the nonsense 39 mutation occurred once (on $++++-$) and has spread to different 5′ haplotypes by recombination in the area between the δ- and β-globin genes. The Sardinian nonsense 39 mutation is unusual in being found on so many haplotypes. On a worldwide scale mutations occur commonly on two, or at most three, seven-site haplotypes; in some cases, such as the 619 bp Indian deletion the association is absolute (Thein *et al.*, 1984). A single haplotype ($+----+$) has been found on more than 40 Asian patients with this deletion.

A few common mutations may have arisen twice, the best examples being the codon 41-TCCT deletion and IVS2:654 mutation. The former occurs on two five-site 3′ haplotypes ($+++-+$ and $--+-+$) which cannot be easily related to each other through conversion and point mutation (Cheng *et al.*, 1984; Kazazian *et al.*, 1986a, 1987; Chan *et al.*, 1987; Zhang *et al.*, 1988; Liu *et al.*, 1988). However, it is possible that a small conversion is responsible for spreading the mutation to different haplotypes.

If, in general, it is true that each mutation arose once, and if the 5′ and 3′ haplotypes freely recombine, then the close association between haplotypes and mutations reflects the relatively recent origin of the mutations. If the mutations were as ancient as the origin of haplotypes, then they would display the same degree of 5′/3′ disequilibrium. Also, because the same association occurs in so many parts of the world, it suggests that β-thalassaemia mutations may have become established at similar times throughout the world. Such a conclusion can only be tentative; the rate of recombination is not constant. It will depend on various features of the structure of the population in which it occurs, for example, the size, the mating pattern and the extent of migration. Were the rate constant, then dating the mutations origins would in theory be feasible.

If it is assumed that the African origin of man is correct, the regional specificity of mutations provides further evidence for their relatively recent origin. Had β-thalassaemia been present at significant frequencies in the founder population, then it is likely that the present inhabitants of

Asia and Europe would share the same mutations, just as they share other genetic markers (for instance the β-globin haplotypes). Of course, it is possible that genetic drift and founder effects could have so disturbed the gene frequencies that they no longer resemble the ancestral pattern, but this is unlikely given that so few mutations are shared. Moreover, when the same mutation is found in Europe and Asia it frequently occurs on different 3' haplotypes.

It is known that β-thalassaemia mutations arise spontaneously in different parts of the world. The appearance of new mutations has been documented; mutations identical with those characterized in the 'malarial belt' have been found in areas devoid of malaria (Tonz *et al.*, 1973; Noronha and Honig, 1978; Stamatoyannopoulos *et al.*, 1981; Kazazian *et al.*, 1986b; Chehab *et al.*, 1986). However, these identical mutations do not occur on identical haplotypes, which suggests their independent origin. It appears, therefore, that in many different parts of the world some factor has elevated the frequency of sporadically occurring mutations, and that this factor acted independently in each locality.

Could malaria be the factor responsible? Is it, in other words, only recently prevalent in Eurasia? At first sight this seems unlikely because there is good reason to believe that malaria has been present since the first appearance of man; the host specificity of the Plasmodia parasites attests to a long period of adaptation. Also, human malaria probably first appeared in Africa (Bruce-Chwatt, 1965).

However, the present day distribution of the disease rules out the possibility that the parasite has been present in all world populations since man first spread out from an ancestral homeland. Malaria is absent from the remote Pacific islands and was only recently introduced into Australia (Bruce-Chwatt, 1965). The main spread of Plasmodia probably therefore occurred after the dispersal of mankind through Eurasia and Oceania.

One likely explanation for the discrepancy between the early origin of malaria and its later spread, is that disease dispersal became possible on a large scale only with the concentration of human populations within the last 30 000 years. Whereas 125 000 years ago the total human population is estimated to have been 125 000, by 25 000 years ago the census number would have been about 3 million. The palaeoepidemiology of malaria can only be a speculative discipline, but its conclusions do fit with the picture derived from the distribution of β-thalassaemia mutations.

The evidence that selection is responsible for the high frequencies of β-thalassaemia is good; the regional specificity of mutations points to local factors rather than founder effects, the tight linkage to haplotype suggests a recent cause (and therefore a powerful one to act so quickly: Nurse, 1985), and the fact that similar observations have been made for β-thalassaemia in so many different parts of the world makes it unlikely that a chance factor, such as random drift, is responsible. The association with malaria is, however, no proof that the disease has a causal role; it is merely that no other factor is known to be prevalent in the areas with high frequencies of β-thalassaemia. The best proof will come from the demonstration that β-thalassaemia red cells in some way protect their human host against malaria.

Conclusion

Studies of genetic variation in human populations can point in two directions. On one hand, knowledge of a population's structure can give information about the molecular mechanisms that give rise to variation. Thus recombination frequency across the β-globin cluster could be estimated by studies in an appropriate population (Chakravarti *et al.*, 1986), and investigation of the number of different mutations occurring in the β-globin gene could identify stretches of DNA that may be more than usually mutation prone (Wong *et al.*, 1986). On the other hand, the distribution of variants can give information about population structure; for instance, the loss of heterozygosity at minisatellite loci in Pacific islanders suggests that that population passed through a bottleneck and has emerged more homogeneous genetically than before (Flint *et al.*, 1989).

However, it is not clear that the two issues can be so easily separated; in attempting to answer questions about population structure (the aim of this chapter), problems soon arise about the nature of the underlying molecular mechanisms responsible for the variants. At the beginning of this chapter it was shown that β- and α-globin haplotypes have a skewed allele distribution. The observation that the distribution is different between Africa and the rest of the world suggested that a population bottleneck might have occurred in the colonization of Europe and Asia. If this is true, then similar skewed distributions should be found for other restriction enzyme polymorphism haplotypes at other loci, and this is indeed the case (Antonarakis *et al.*, 1988b; Woo, 1988; Leitersdorf *et al.*, 1989; Daiger *et al.*, 1989a, 1989b). But what remains strange is how little the β-globin haplotype frequencies vary across the world. It is almost impossible to separate Vietnamese from Portuguese. Surely bottlenecks must have occurred elsewhere in the colonization of Asia. A further puzzle is that even though the Africans have a different allele distribution, this is still skewed, just like the European distribution. These peculiarities suggest that more needs to be known about the molecular mechanisms governing recombination in order to explain haplotype distributions.

A second example of the way in which knowledge of molecular mechanisms and population structure become inextricably interwoven arises from the study of α^+-thalassaemia. Clearly selection has elevated its frequencies in world populations, but why has selection not pushed it to fixation? The effect of the gene on haematological indices is so slight that the heterozygote is almost indistinguishable from normal. What is keeping the gene at frequencies of only 80%, and why is there such a marked cline in the Pacific? Assuming that it has such a benign phenotype, there seems no reason for its frequency to remain so low in those areas where malaria is prevalent, even when malarial endemicity may not be high. Perhaps the answer is to be found in the pathophysiology of the disease. It may be that we simply do not know enough about α-globin gene function.

Nevertheless we can be certain that genetic drift, founder effect and selection are responsible for genetic variation in human populations and we have seen how complex a picture they can produce. What needs to be emphasized, however, is that although these processes have been demonstrated using data from the new molecular technology, the results are also dependent on well established principles of population genetics. Much of the work described in this chapter has been on remote, 'low-technology' populations, whose genetic structure may seem to have little relation to the populations of the developed West. Yet it has been clear for some time that to understand the factors that maintain genetic variation one must turn to those few populations so far unaffected by the social, economic and technological changes that characterize 'developed' countries. Studying the peoples of the 'developed' world is unlikely to be informative because, in genetic terms, they are still coalitions of tribal isolates (Chakraborty *et al.*, 1988); not enough time has elapsed for the changes in mating patterns and migration that accompany urbanization and long distance travel to transform the genetic make-up of these populations; they are not yet in equilibrium.

Biological anthropologists have looked hard for populations that may help to characterize the population structure possessed by humankind for most of its evolutionary history. Neel and his associates penetrated the Brazilian jungle in this endeavour (Neel, 1978), Morton the Pacific (Imaizumi and Morton, 1970), and Cavalli-Sforza the interior of Africa

162   J. Flint, J. B. Clegg and A. J. Boyce

(Cavalli-Sforza, 1986). In a similar spirit of scientific enterprise the inhabitants of Papua New Guinea have been repeatedly venesected over the few years in which they have made contact with Europeans.

The globin gene studies described here have confirmed the importance of such investigations. To show how selection alters gene frequencies in populations it was necessary to investigate the inhabitants of Melanesia, and work on the Polynesians has indicated how the genetic structure of the present is shaped by the events of the past. Analysis of the distribution of β-globin haplotypes makes clear that similar forces can still be traced in the genetic make-up of Europeans; their restricted set of β-globin haplotypes is a bequest from the original settlers of Europe. Although the technology has been revolutionized, the epidemiological approaches to the problems of population structure remain the same.

References
Akar, N., Cavar, A. O., Dessi, E., Loi, A., Pirastu, M. and Cao, A. (1987). β thalassaemia mutations in the Turkish population. *Journal of Medical Genetics*, **24**, 378–81.
Allen, J., Gosden C., Jones, R. and White, P. J. (1988). Pleistocene dates for the human occupation of New Ireland, Northern Melanesia. *Nature*, **331**, 707–9.
Allison, A. C. (1964). Polymorphism and natural selection in human populations. *Cold Spring Harbor Symposia on Quantitative Biology*, **29**, 137–49.
Amselem, S., Nunes, V., Vidaud, M., Estivill, X., Wong, C., d'Auriol, L., Vidaud, D., Galibert, F., Baiget, M. and Goossens, M. (1988). Determination of the spectrum of β thalassaemia genes in Spain by dot-blot analysis of amplified β globin DNA. *American Journal of Human Genetics*, **43**, 95–100.
Anagnou, N. P., O'Brien, J. J., Shimada, T., Nash, W. G., Chen, M.-J. and Nienhuis, A. W. (1984). Chromosomal organization of the human dihydrofolate reductase genes: dispersion, selective amplification and a novel form of polymorphism. *Proceedings of the National Academy of Sciences USA*, **81**, 5170–4.
Antonarakis, S. E., Boehm, C. D., Giardina, P. J. C. and Kazazian, H. H. (1982b). Non-random association of polymorphic restriction sites in the β globin gene cluster. *Proceedings of the National Academy of Sciences USA*, **79**, 137–41.
Antonarakis, S. E., Boehm, C. D., Serjeant, G. R., Theisen, C. E., Dover, G. J. and Kazazian, H. H. (1984a). Origin of the β^S-globin gene in Blacks: the contribution of recurrent mutation or gene conversion or both. *Proceedings of the National Academy of Sciences USA*, **81**, 853–6.
Antonarakis, S. E., Kang, J., Lam, V. M. S., Tam, J. W. O. and Li, A. M. C. (1988a). Molecular characterization of β globin gene mutations in patients with β thalassaemia intermedia in South China. *British Journal of Haematology*, **70**, 357–61.

Antonarakis, S. E., Kazazian, H. H. and Orkin, S. H. (1985). DNA polymorphisms and molecular pathology of the human globin gene clusters. *Human Genetics*, **69**, 1–14.

Antonarakis, S. E., Oettgen, P., Chakravarti, A., Halloran, S. L., Hudson, R. R., Feisee, L. and Karathanasis, S. K. (1988b). DNA polymorphism haplotypes of the human apolipoprotein APOA1-APOC3-APOA4 gene cluster. *Human Genetics*, **80**, 265–73.

Antonarakis, S. E., Orkin, S. H., Cheng, T.-C., Scott, A. F., Sexton, J. P., Trusko, S. P., Charache, S. and Kazazian H. H. (1984b). β thalassaemia in American blacks: novel mutation in the 'TATA' box and an acceptor splice site. *Proceedings of the National Academy of Sciences USA*, **81**, 1154–8.

Antonarakis, S. E., Orkin, S. H., Kazazian, H. H., Goff, S. C., Boehm, C. D., Waber, P. G., Sexton, J. P., Ostrer, H., Fairbanks, V. F. and Chakravarti, A. (1982a). Evidence for multiple origins of the β^E globin gene in Southeast Asia. *Proceedings of the National Academy of Sciences USA*, **79**, 6608–11.

Athanassiadou, A., Zarkadis, I., Papahadjopoulou, A. and Maniatis, G. M. (1987). DNA haplotype heterogeneity of β thalassaemia in Greece: feasibility of prenatal diagnosis. *British Journal of Haematology*, **66**, 379–83.

Aulehla-Scholz, C., Basaran, S., Agaoglu, L., Arcasoy, A., Holzgreve, W., Miny, P., Ridolfi, F. and Horst, J. (1990). Molecular basis of β thalassaemia in Turkey: detection of rare mutations by direct sequencing. *Human Genetics*, **84**, 195–7.

Bell, G. I., Horita, S. and Karam, J. H. (1984). A polymorphic locus near the human insulin gene is associated with insulin-dependent diabetes mellitus. *Diabetes*, **33**, 176–84.

Bellwood, P. S. (1989). The colonization of the Pacific: some current hypotheses. In *The Colonization of the Pacific: a genetic trail*, ed. A. V. S. Hill and S. W. Serjeantson, pp. 1–59. Oxford: Oxford University Press.

Black, R. H. (1954). *Some aspects of malaria in the New Hebrides*. South Pacific Commission Technical Paper No. 60.

Boehm, C. D., Dowling, C. E., Antonarakis, S. E., Honig, G. R. and Kazazian, H. H. (1985). Evidence supporting a single origin of the β^C-globin gene in Blacks. *American Journal of Human Genetics*, **37**, 771–7.

Borts, R. H. and Haber, J. E. (1990). Length and distribution of meiotic gene conversion tracts and crossovers in *Saccharomyces cervisiae*. *Genetics*, **123**, 69–80.

Bowcock, A. M., Bucci, C., Hebert, J. M., Kidd, J. R., Kidd, K. K., Friedlander, J. S. and Cavalli-Sforza, L. L. (1987). Study of 47 DNA markers in five populations from four continents. *Gene Geography*, **1**, 47–64.

Bowden, D. K., Hill, A. V. S., Higgs, D. R., Oppenheimer, S. J., Weatherall, D. J. and Clegg, J. B. (1987). Different hematologic phenotypes are associated with leftward ($-\alpha^{4.2}$) and rightward ($-\alpha^{3.7}$) α thalassaemia deletions. *Journal of Clinical Investigation*, **79**, 39–43.

Bowden, D. K., Hill, A. V. S., Higgs, D. R., Weatherall, D. J. and Clegg, J. B. (1985). The relative roles of genetic factors, dietary deficiency and infection in anaemia in Vanuatu, Southwest Pacific. *Lancet*, **ii**, 1025–8.

164   J. Flint, J. B. Clegg and A. J. Boyce

Bowman, J. E. (ed.) (1983). *Distribution and Evolution of the Hemoglobin and Globin Loci*. New York: Elsevier Science Publications.

Brockway, R. W. (1983). The origin and dispersal of the Polynesians: some recent evidence. *Journal of Human Biology*, **12**, 501–3.

Brown, P. J. (1981). New considerations on the distribution of malaria, thalassaemia and glucose–6-phosphate dehydrogenase deficiency in Sardinia. *Human Biology*, **53**, 367–82.

Bruce-Chwatt, L. J. (1965). Paleogenesis and paleo-epidemiology of primate malaria. *Bulletin of the World Health Organization*, **32**, 363–87.

Buck, P. H. (1938). *Vikings of the Sunrise*. York: Frederick A. Stokes.

Bunn, H. F. and Forget, B. G. (1986). *Hemoglobin: Molecular Genetic and Clinical Aspects*. Philadelphia: Saunders.

Camaschella, C., Saglio, G., Serra, A., Guerrasio, A., Bertero, T., Rege-Cambrin, G., Loi, A. and Pirastu, M. (1988a). Molecular characterization of thalassaemia intermedia in Italy. *Birth Defects*, **23**, 111–16.

Camaschella, C., Serra, A., Saglio, G., Bertero, M. T., Mazza, U., Terzoli, S., Brambati, B., Cremonesi, L., Travi, M. and Ferrari, M. (1988b). Meiotic recombination in the β globin cluster causing an error in prenatal diagnosis of β thalassaemia. *Journal of Medical Genetics*, **25**, 307–10.

Cann, R. L., Stoneking, M. and Wilson, A. C. (1987). Mitochondrial DNA and human evolution. *Nature*, **325**, 31–6.

Carestia, C., Pagano, L., Fioretti, G. and Mastrobuoni, A. (1987). β thalassaemia in Campania: DNA polymorphism analysis in β^A and β^{thal} chromosomes and its usefulness in prenatal diagnosis. *British Journal of Haematology*, **67**, 231–4.

Cavalli-Sforza, L. L. (ed.) (1986). *Research on Pygmies*. Orlando: Academic Press.

Cavalli-Sforza, L. L., Piazza, A., Menozzi, P. and Mountain, J. (1988). Reconstruction of human evolution: bringing together genetic, archaeological and linguistic data. *Proceedings of the National Academy of Sciences USA*, **85**, 6002–6.

Chakraborty, R., Smouse, P. E. and Meel, J. V. (1988). Population amalgamation and genetic variation: observations on artificially agglomerated tribal populations of Central and South America. *American Journal of Human Genetics*, **43**, 709–25.

Chakravarti, A., Buetow, K. H., Antonarakis, S. E., Waber, P. G., Boehm, C. D. and Kazazian, H. H. (1986). Nonuniform recombination within the human β globin gene cluster. *American Journal of Human Genetics*, **36**, 1239–58.

Chakravarti, A., Phillips, J. A., Mellitus, K. H., Buetow, K. H. and Seeburgh, P. H. (1984). Patterns of polymorphism and linkage disequilibrium suggest independent origins of the human growth hormone gene cluster. *Proceedings of the National Academy of Sciences USA*, **88**, 6085–9.

Chan, V., Chan, T. K., Leung, N. K., Kan, Y. W. and Todd, D. (1986). Characteristics and distribution of β thalassaemia haplotypes in South China. *Human Genetics*, **73**, 23–6.

Chan, V., Chan, T. K., Chehab, F. F. and Todd, D. (1987). Distribution of β thalassaemia mutations in South China and their association with haplotypes. *American Journal of Human Genetics*, **41**, 678–85.

Chebloune, Y., Pagnier, J., Trabuchet, G., Faure, C., Verdier, G., Labie, D. and Nigon, V. (1988). Structural analysis of the 5' flanking region of the β globin gene in African sickle cell anemia patients: further evidence for three origins of the sickle mutation in Africa. *Proceedings of the National Academy of Sciences USA*, **85**, 4431–5.

Chehab, F. F., Der Kaloustian, V., Khouri, F. P., Deeb, S. S. and Kan, Y. W. (1987). The molecular basis of β thalassaemia in Lebanon: application to prenatal diagnosis. *Blood*, **69**, 1141–5.

Chehab, F. F., Honig, G. R. and Kan, Y. W. (1986). Spontaneous mutation in β thalassaemia producing the same nucleotide substitution as that in a common hereditary form. *Lancet*, **i**, 3–5.

Cheng, T.-C, Orkin, S. H., Antonarakis, S. E., Potter, M. J., Sexton, J. P., Markham, A. F. Giardina, P. J. V., Li, A. and Kazazian, H. H. (1984). β thalassaemia in Chinese: Use in *in vivo* RNA analysis and oligonucleotide hybridization in systematic characterization of molecular defects. *Proceedings of the National Academy of Sciences USA*. **81**, 2821–5.

Chibani, J., Vidard, M., Duquesnoy, P., Berge-Lefranc, J. L., Pirastu, M., Ellouze, R., Rosa, J. and Goossens, M. (1988). The peculiar spectrum of β thalassaemia genes in Tunisia. *Human Genetics*, **78**, 190–2.

Clarke, R. (1979). Language. In *The Prehistory of Polynesia*, ed. J. D. Jennings, pp. 249–70. Cambridge, MA: Harvard University Press.

Clegg, J. B. (1987). Can the product of the θ-gene be a real globin? *Nature*, **329**, 465–7.

Conner, B. J., Reyes, A. A., Morin, C., Itakura, K., Teplitz, R. L. and Wallace, R. B. (1983). Detection of sickle cell $β^S$-globin allele by hybridization with synthetic oligonucleotides. *Proceedings of the National Academy of Sciences USA*, **80**, 278–82.

Coutinho-Gomes, M. P., Gomes da Costa, M. G., Braga, L. B., Cordeiro-Ferreira, N. T., Loi, A., Pirastu, M. and Cao, A. (1988). β thalassaemia mutations in the Portuguese population. *Human Genetics*, **78**, 13–15.

Daiger, S. P., Chakraborty, R., Reed, L., Fokete, G., Schuler, D., Berenssi, G., Nazz, I., Brdicka, R., Kamaryt, J., Pijackova, A., Moore, S., Sullivan, S. and Woo, S. L. C. (1989a). Polymorphic DNA haplotypes at the phenylalanine hydroxylase (PAH) locus in European families with phenylketonuria (PKU). *American Journal of Human Genetics*, **45**, 310–18.

Daiger, S. P., Reed, L., Huang, S.-S, Zeny, Y.-T, Wang, T., Lo, W. H. Y., Okano, Y., Hase, Y., Fukuda, Y., Oura, T., Tada, K. and Woo, S. L. C. (1989b). Polymorphic RNA haplotypes at the phenylalanine hydroxylase (PAH) locus in Asian families with phenylketonuria (PKU). *American Journal of Human Genetics*, **45**, 319–24.

Del Senno, L., Pirastu, M., Barbieri, R., Bernardi, F., Buzzoni, D., Manchietti, G., Perrotta, C., Vullo, C., Kan, Y. W. and Conconi, F. (1985). β thalassaemia in the Po river delta region (northern Italy): genotype and β globin synthesis. *Journal of Medical Genetics*, **22**, 54–8.

Deisseroth, A., Nienhuis, A., Turner, P., Velez, R., Anderson, W. F., Ruddle, F., Lawrence, J., Creagan, R. and Kucherlapati, R. S. (1977). Localization of the human α globin structural gene to chromosome 16 in somatic cell hybrids by molecular hybridization assay. *Cell*, **12**, 205–18.

Deisseroth, A., Nienhuis, A., Lawrence, J., Giles, R., Turner, P. and Ruddle, F. H. (1978). Chromosomal localization of human β globin gene on human chromosome 11 in somatic cell hybrids. *Proceedings of the National Academy of Sciences USA*, **75**, 1456–60.

Diamond, J. M. and Rotter, J. I. (1987). Observing the founder effect in human evolution. *Nature*, **329**, 105–6.

Diaz-Chico, T. C., Yang, K. G., Stoming, T. A., Efremov, D. G., Kutlar, A., Kutlar, F., Aksoy, M., Altay, C., Gurgey, A., Kilinc, Y. and Huisman, T. H. J. (1988). Mild and severe β thalassaemia among homozygotes in Turkey: Identification of the types by hybridization of amplified DNA with synthetic probes. *Blood*, **71**, 248–51.

Di Marzo, R., Dowling, C. E., Wong, C., Maggio, A. and Kazazian, H. H. (1988). Spectrum of β thalassaemia mutations in Sicily. *British Journal of Haematology*, **69**, 393–7.

Edwards, A. W. F. (1986). Evolutionary relationship of human populations. *Nature*, **323**, 744.

Efstratiadis, A., Posakony, J. W., Maniatis, T., Lawn, R. M., O'Connell, C., Spritz, R. A., DeRiel, J.-K., Forget, B. G., Weissman, S. M., Slightom, J. L., Blechl, A. E., Smithes, O., Baralle, F. E., Shoulders, C. C. and Proudfoot, N. J. (1980). The structure and evolution of the human β globin gene family. *Cell*, **21**, 653–68.

Embury, S. H. (1985). The interaction of co-existent α thalassaemia and sickle cell anemia: a model for the clinical and cellular results of diminished polymerization? *Annals of the New York Academy of Science*, **445**, 37–44.

Embury, S. H., Miller, J. A., Dozy, A. M., Kan, Y. W., Chan, V. and Todd, D. (1980). Two different molecular organizations account for the single α globin gene of the α thalassaemia-2 genotype. *Journal of Clinical Investigation*, **66**, 1319–25.

Fermi, C. (1934). *Regioni malariche, decadenza risanamento E spesa 'Sardegna'. Vol.I, Sassani Province*. Rome: Tipografia dello Stato.

  (1938). *Regioni malariche, decadenza risanamento E spesa 'Sardegna'. Vol. 2, Nueovo Province; Vol. 3, Cagliani Province*. Rome: Tipografia dello Stato.

Flint, J., Boyce, A. J., Martinson, J. J. and Clegg, J. B. (1989). Population bottlenecks in Polynesia revealed by minisatellites. *Human Genetics*, **83**, 252–63.

Flint, J., Hill, A. V. S., Bowden, D. K., Oppenheimer, S. J., Sill, P. R., Serjeantson, S. W., Bana-Koiri, J., Bhatia, K., Alpers, M. P., Boyce, A. J., Weatherall, D. J. and Clegg, J. B. (1986). High frequencies of α thalassaemia are the result of natural selection by malaria. *Nature*, **321**, 744–9.

Fucharoen, S., Fucharoen, G., Sriroongueng, W., Laosombat, V., Jetsrisuparb, A., Prasatkaew, S., Tanphaichtr, U. S., Suvatte, V., Tuchinda, S. and Fukumaki, Y. (1989). Molecular basis of β thalassaemia in Thailand: analysis of β thalassaemia mutations using the polymerase chain reaction. *Human Genetics*, **84**, 41–6.

Fukumaki, Y., Matsunaga, E., Takihara, Y., Nakamura, T., Takagi, Y., Tan-phaichitr, V. S., Suvatte, V., Tuchinda, S., Lin, S.-T. and Lee, H.-T. (1988). Multiple origins of the β thalassaemia gene with a four nucleotide deletion in its second exon. *Birth Defects*, **23**, 81–5.

Gerhard, D. S., Kidd, K. K., Kidd, J. R., Egeland, J. A. and Houseman, D. E. (1984). Identification of a recent recombination event with in the human β globin gene cluster. *Proceedings of the National Academy of Sciences USA*, **81**, 7875–9.

Giampaolo, A., Mavilio, F., Massa, A., Gabbianelli, M., Guerriero, R., Sposi, N. N., Care, A., Cianciulli, P., Tentori, L. and Marinucci, M. (1984). Molecular heterogeneity of beta thalassaemia in the Italian population. *British Journal of Haematology*, **56**, 79–85.

Giles, E. and Ambrose, S. H. (1986). Are we all out of Africa? *Nature*, **322**, 21–2.

Giles, E., Wybar, S. and Walsh, R. J. (1970). Microevolution in New Guinea: additional evidence for genetic drift. *Archaeology and Physical Anthropology of Oceania*, **5**, 60–72.

Goodbourn, S. E. Y., Higgs, D. R., Clegg, J. B. and Weatherall, D. J. (1983). Molecular basis of length polymorphism in the human ζ-globin gene complex. *Proceedings of the National Academy of Sciences USA*, **80**, 5022–6.

Goossens, M., Dozy, A. M., Embury, S. H., Zachariades, Z., Hadjiminas, M. G., Stamatoyannopoulos, G. and Kan, Y. W. (1980). Triplicated α globin loci in humans. *Proceedings of the National Academy of Sciences USA*, **77**, 518–21.

Green, R. C. (1979). Lapita. In *The Prehistory of Polynesia*, ed. J. D. Jennings, pp. 22–60. Cambridge, MA: Harvard University Press.

Groube, L., Chappell, J., Muke, J. and Price, D. (1986). A 40 000 year old human occupation site at Huon Peninsula, Papua New Guinea. *Nature*, **324**, 453–5.

Haldane, J. B. S. (1948). The rate of mutation of human genes. *Proceedings of the 8th International Congress on Genetics, Hereditas Suppl.* **35**, pp. 267–73.

Hararo, T., Reese, A. L., Ryan, R., Abraham, B. L. and Huisman, T. H. J. (1985). Five haplotypes in black β thalassaemia heterozygotes: three are associated with high and two with low $^{G}\gamma$ values in HbF. *British Journal of Haematology*, **59**, 333–42.

Hertzberg, M., Jahromi, K., Ferguson, V., Dahl, H. H. M., Mercer, J., Mickleson, K. N. P. and Trent, R. J. (1989a). Phenylalanine hydroxylase gene haplotypes in Polynesians. Evolutionary origins and absence of alleles associated with severe phenylketonuria. *American Journal of Human Genetics*, **44**, 382–7.

Hertzberg, M. S., Mickleson, K. N. P., Serjeantson, S. W., Prior, J. F. and Trent, R. J. (1989b). An Asian specific 9 bp deletion is frequently found in Polynesians. *American Journal of Human Genetics*, **44**, 504–10.

Hertzberg, M. S., Mickleson, K. N. P. and Trent, R. J. (1988). α globin gene haplotypes in Polynesians: their relationships to population groups and gene rearrangements. *American Journal of Human Genetics*, **43**, 971–7.

Hess, J. F., Schmid, C. W., Shen and C.-K. J. (1984). A gradient of sequence divergence in the human adult α globin duplication units. *Science*, **226**, 67–70.

Higgs, D. R., Aldridge, B. E., Lamb, J., Clegg, J. B., Weatherall, D. J., Hayes, R. J., Grandison, Y., Lowrie, Y., Mason, K. P., Serjeant, B. E. and Serjeant, G. R. (1982). The interaction of alpha thalassaemia and homozygous sickle-cell disease. *New England Journal of Medicine*, **306**, 1441–6.

Higgs, D. R., Hill, A. V. S., Bowden, D. K., Weatherall, D. J. and Clegg, J. B. (1984). Independent recombination events between the duplicated human α globin genes; implications for their concerted evolution. *Nucleic Acids Research*, **12**, 6965–77.

Higgs, D. R., Old, J. M., Pressley, L., Clegg, J. B. and Weatherall, D. J. (1980). A novel α globin gene arrangement in man. *Nature*, **284**, 632–5.

Higgs, D. R., Vickers, M. A., Wilkie, A. O. M., Pretorius, I.-M., Jarman, A. P. and Weatherall, D. J. (1989). A review of the molecular genetics of the human α globin gene cluster. *Blood*, **73**, 1081–104.

Higgs, D. R., Wainscoat, J. S., Flint, J., Hill, A. V. S., Thein, S. L., Nicholls, R. D., Teal, H., Ayyub, H., Peto, T. E. A., Falusi, A. G., Jarman, A. P., Clegg, J. B. and Weatherall, D. J. (1986). Analysis of the human α globin gene cluster reveals a highly informative genetic locus. *Proceedings of the National Academy of Sciences USA*, **83**, 5165–9.

Hill, A. V. S. and Serjeantson, S. W. (ed.) (1989). *The Colonization of the Pacific: A Genetic Trail*. Oxford: Oxford University Press.

Hill, A. V. S., Bowden, D. K., Trent, R. J., Higgs, D. R., Oppenheimer, S. J., Thein, S. L., Mickleson, K. N. P., Weatherall, D. J. and Clegg, J. B. (1985a). Melanesians and Polynesians share a unique α thalassaemia mutation. *American Journal of Human Genetics*, **37**, 571–80.

Hill, A. V. S., Bowden, D. K., Weatherall, D. J. and Clegg, J. B. (1986). Chromosomes with one, two, three and four fetal globin genes: molecular and hematologic analysis. *Blood*, **67**, 1611–18.

Hill, A. V. S., Gentile, B., Bonnardot, J. M., Roux, J. K., Weatherall, D. J. and Clegg, J. B. (1987). Polynesian origins and affinities: globin gene variants in Eastern Polynesia. *American Journal of Human Genetics*, **40**, 453–63.

Hill, A. V. S., Nicholls, R. D., Thein, S. L. and Higgs, D. R. (1985b) Recombination within the human embryonic ζ globin locus: a common ζ-ζ chromosome produced by gene conversion of the ζ gene. *Cell*, **42**, 809–19.

Hill, A. V. S., O'Shaughnessy, D. R. and Clegg, J. B. (1989). Haemoglobin and globin gene variants in the Pacific. In *The Colonization of the Pacific: A Genetic Trail*, ed. A. V. S. Hill and S. W. Serjeantson, pp. 246–85. Oxford: Oxford University Press.

Hill, W. G. and Weir, B. S. (1988). Variances and co-variances of squared linkage disequilibria in finite populations. *Theoretical Population Biology*, **33**, 54–78.

Howells, W. W. (1973). *The Pacific Islanders*. New York: Scribners and Sons. (1976). Explaining modern man: evolutionists versus migrationists. *Journal of Human Evolution*, **5**, 477–96.

Hsu, S.-L., Marks, J., Shaw, J.-P., Tam, M., Higgs, D. R., Shen, C. C. and Shen C.-K. J. (1988). Structure and expression of the human θ_1-globin gene. *Nature*, **331**, 94–6.

Huang, S.-Z., Kazazian, H. H., Waber, P. G., Lo, W. H. Y., Cai, R.-L. and Wang, M. Q. (1985). β thalassaemia in Chinese: analysis of polymorphic restriction site haplotypes in the β globin gene cluster. *Chinese Medical Journal*, **98**, 881–6.

Huang, S.-Z., Wong, C., Antonarakis, S. E., Ro-lein, T., Lo, W. H. Y. and Kazazian, H. H. (1986). The same 'TATA' box β thalassaemia mutation in Chinese and US blacks: another example of independent origins of mutation. *Human Genetics*, **74**, 162–4.

Huang, S.-Z., Zhou, X.-D., Zhu, H., Ren, Z.-R. and Zeng, X.-T. (1990). Detection of β thalassaemia mutations in the Chinese using amplified DNA from dried blood specimens. *Human Genetics*, **84**, 129–31.

Hundrieser, J., Sanguansermsri, T., Papp, T., Laig, M. and Flatz, G. (1988). β globin gene linked DNA haplotypes and frameworks in three south-east Asian populations. *Human Genetics*, **80**, 90–4.

Imaizumi, Y. and Morton, N. E. (1970). Isolation by distance in New Guinea and Micronesia. *Archaeology and Physical Anthropology of Oceania*, **5**, 218–35.

Jarman, A. P. and Higgs, D. R. (1988). A new hypervariable marker of the human α globin cluster. *American Journal of Human Genetics*, **42**, 8–16.

Jarman, A. P., Nicholls, R. D., Weatherall, D. J., Clegg, J. B. and Higgs, D. R. (1986). Molecular characterization of a hypervariable region downstream of the human α globin gene cluster. *EMBO Journal*, **5**, 1857–63.

Jeffreys, A. J., Wilson, V. and Thein, S. L. (1985). Hypervariable 'minisatellite' regions in human DNA. *Nature*, **314**, 67–73.

Jennings, J. D. (ed.) (1979). *The Prehistory of Polynesia*. Cambridge, MA: Harvard University Press.

Jones, J. S. (1986). The origin of *Homo sapiens*: the genetic evidence. In *Major Trends in Primate and Human Evolution*, ed. B. Wood, L. Martin and P. Andrews, pp. 317–330. Cambridge: Cambridge University Press.

Jones, J. S. and Rouhani, S. (1986a). Human evolution: how small was the bottleneck? *Nature*, **319**, 449–50.

(1986b). Mankind's genetic bottleneck. *Nature*, **322**, 599–600.

Jonxis, J. H. P. (ed.) (1965). *Abnormal Haemoglobins in Africa*. Philadelphia: F. A. Davis.

Kan, Y. W. and Dozy, A. M. (1978). Polymorphism of DNA sequence adjacent to human β globin structural gene: relationship to sickle mutation. *Proceedings of the National Academy of Sciences USA*, **75**, 5631–5.

(1980). Evolution of the hemoglobin S and C genes in World populations. *Science*, **209**, 492–3.

Kaplan, F., Kokotsis, G., De Braekeleer, M., Morgan, K. and Scriver, C. R. (1990). β thalassaemia genes in French-Canadians; haplotype and mutation analysis of Portneuf chromosomes. *American Journal of Human Genetics*, **46**, 126–32.

Kazazian, H. H. and Antonarakis, S. E. (1988). The varieties of mutation. *Progress in Medical Genetics*, **7**, 43–67.

Kazazian, H. H. and Boehm, C. D. (1988). Molecular basis and prenatal diagnosis of β thalassaemia. *Blood*, **72**, 1107–16.

Kazazian, H. H., Dowling, C. E., Waber, P. S., Huang, S. and Lo, W. H. Y. (1986a). The spectrum of β thalassaemia genes in China and Southeast Asia. *Blood*, **68**, 964–6.

170   J. Flint, J. B. Clegg and A. J. Boyce

Kazazian, H. H., Dowling, C. E., Waber, P. G., Huang, S.-Z., Lo, W. H. Y., Li, A., Jam, H. W. O., Kang, K. and Antonarakis, S. E. (1987). Molecular characterization of β thalassaemia major and β thalassaemia intermedia in China and Southeast Asia. In *Developmental Control of Globin Gene Expression*, ed. G. Stamatoyannopoulos and A. W. Nienhuis, pp. 401–12. New York: Alan R. Liss.

Kazazian, H. H., Orkin, S. J., Antonarakis, S. E., Sexton, J. P., Boehm, C. D., Goff, S. C. and Waber, P. G. (1984a). Molecular characterization of seven β thalassaemia mutations in Asian Indians. *EMBO Journal*, 3, 593–6.

Kazazian, H. H., Orkin, S. H., Boehm, C. D., Goff, S. C., Wong, C., Dowling, C. E., Newburger, P. E., Knowlton, R. G., Brown, V. and Donis-Keller, H. (1986b). Characterization of a spontaneous mutation to a β thalassaemia allele. *American Journal of Human Genetics*, 38, 860–7.

Kazazian, H. H., Orkin, S. J., Markham, A. F., Chapman, C. R., Youssoufian, H. and Waber, P. G. (1984b). Quantification of the close association between DNA haplotypes and specific β thalassaemia mutations in Mediterraneans. *Nature*, 310, 152–4.

Kazazian, H. H., Waber, P. G., Boehm, C. D., Lee, J. I., Antonarakis, S. E. and Fairbank, V. F. (1984c). Hemoglobin E in Europeans: further evidence for multiple origins of the β^E globin gene. *American Journal of Human Genetics*, 36, 212–17.

Kimura, A., Matsunaga, E., Ohta, Y., Fujiyoshi, T., Matsuo, T., Nakamura, T., Imamura, T., Yanase, T. and Takagi, Y. (1983). Structure of cloned delta globin genes from a normal subject and a patient with delta-thalassaemia: sequence polymorphisms found in the delta globin gene region of Japanese individuals. *Nucleic Acids Research*, 10, 5625–732.

Kirch, P. V. (1986). Rethinking east Polynesian prehistory. *Journal of Polynesian Society*, 95, 9–40.

Kirk, R. L. (1979). Genetic differentiation in Australia and the Western Pacific and its leaning on the origin of the first Americans. In *The First Americans: Origins, Affinities and Adaptations*, ed. W. S. Laughlin, pp. 211–37. New York: Gustav Fisher.

(1989). Population genetic studies in the Pacific: red cell antigens, serum protein and enzyme systems. In *The Colonization of the Pacific: A Genetic Trail*, ed. A. V. S. Hill and S. W. Serjeantson, pp. 60–119. Oxford: Oxford University Press.

Kulozik, A. E., Kar, B. C., Serjeant, G. R., Serjeant, B. E. and Weatherall, D. J. (1988). The molecular basis of α thalassaemia in India: its interaction with the sickle gene. *Blood*, 71, 467–72.

Kulozik, A. E., Wainscoat, J. S., Serjeant, G. R., Al-Awamy, B., Essan, F., Falusi, A.-G., Haque, S. K., Hilali, A. M., Kate, S., Ranasinghe, W. A. C. P. and Weatherall, D. J. (1986). Geographical survey of β^S-globin gene haplotypes: evidence for an independent Asian origin of the sickle cell mutation. *American Journal of Human Genetics*, 39, 239–44.

Laig, M., Sanguansermsi, T., Wiangnon, S., Hundrieser, J., Pape, M. and Flatz, G. (1989). The spectrum of β thalassaemia mutations in northern and northeastern Thailand. *Human Genetics*, 84, 47–50.

Lambert, S. M. (1949). Malaria incidence in Australia and the South Pacific. In *Malariology*, ed. M. F. Boyd, pp. 820–30. Philadelphia: W. B. Saunders.

Lauer, J., Shen, C.-K. J. and Maniatis, T. (1980). The chromosomal arrangement of human α-like globin genes – sequence homology and α globin gene deletions. *Cell*, **20**, 119–30.

Lawn, R. M., Efstratiadis, A., O'Connell, C. and Maniatis, T. (1980). The nucleotide sequence of the human beta globin gene. *Cell*, **21**, 647–51.

Leitersdorf, E., Chakravarti, A. and Hobbs, H. H. (1989). Polymorphic DNA haplotypes at the LDL receptor locus. *American Journal of Human Genetics*, **44**, 409–21.

Lie-Injo, L. E., Cai, S.-P., Wahidisat, I., Moeslichan, S., Lim, L. M., Evangelista, L., Doherty, M. and Kan, Y. W. (1989). β thalassaemia mutations in Indonesia and their linkage to β haplotypes. *American Journal of Human Genetics*, **45**, 971–5.

Lie-Injo, L. E., Herrera, A. R. and Kan, Y. W. (1981). Two types of triplicated α globin loci in humans. *Nucleic Acids Research*, **9**, 3707–17.

Lie-Injo, L. E., Pawson, I. G. and Solair, A. (1985). High frequency of triplicated α globin loci and absence or low frequency of α thalassaemia in Polynesian Samoans. *Human Genetics*, **70**, 116–18.

Litt, M. and Luty, J. A. (1989). A hypervariable microsatellite revealed by *in vitro* amplification of a dinucleotide repeat within the cardiac muscle actin gene. *American Journal of Human Genetics*, **44**, 397–401.

Liu, V. W. S., Woo, Y. K., Lam, V. M. S., Huang, C. H., Chan, A. S., Lam, S. T. S., Wong, H. W. and Tam, J. W. O. (1988). Molecular studies of β thalassaemia DNA of Chinese patients. *Birth Defects*, **23**, 87–92.

Livingstone, F. B. (1985). *Frequencies of Haemoglobin Variants*. Oxford, Oxford University Press.

Logan, J. A. (1953). *The Sardinian Project – an Experiment in the Eradication of the Indigenous Malarious Vector*. Baltimore, MD: Johns Hopkins Press.

Long, J. C., Chakravarti, A., Boehm, C. D., Antonarakis, S. and Kazazian, H. H. (1990). Phylogeny of human β globin haplotypes and its implications for recent human evolution. *American Journal of Physical Anthropology*, **81**, 113–30.

Lucotte, G., Guerin, P., Halle, L., Loirat, F. and Hazout, S. (1989). Y chromosome DNA polymorphisms in two African populations. *American Journal of Human Genetics*, **45**, 16–20.

Lynch, J., Tate, V. E., Weatherall, D. J., Fucharoen, S., Tanphaichitr, V. S., Isarangkura, P., Seksam, P., Laosombat, V., Kulapongs, P. and Wasi, P. (1988). Molecular basis of β thalassaemia in Thailand. *Birth Defects*, **23**, 71–9.

Maggio, A., Acuto, S., DiMarzo, R., LoGioco, P., Giambona, A., Sammarco, P., Siciliano, S. and Caronia, F. (1988). β thalassaemia mutations in Sicily. *Birth Defects*, **23**, 107–10.

Maggio, A., Acuto, S., LoGioco, P., DiMarzo, R., Giambona, A., Sammarco, P. and Caronia, F. (1986). β^A and β^{thal} DNA haplotypes in Sicily. *Human Genetics*, **72**, 229–30.

Mellars, P. and Stringer, C. B. (ed.) (1989). *The Origins and Dispersal of Modern Humans: Behavioural and Biological Perspectives*. Edinburgh: Edinburgh University Press.

Michelson, A. M. and Orkin, S. H. (1983). Boundaries of gene conversion within the duplicated human α globin genes. Concerted evolution by segmental recombination. *Journal of Biological Chemistry*, **258**, 15245–54.

Millard, M., Borge-Lefranc, J. L., Leva, D. and Cartouzon, G. (1987). Oligonucleotide screening of β thalassaemia mutations in the southeast of France. *Hemoglobin*, **11**, 317–27.

Monteiro, C., Rueff, J., Falcao, A. B., Portugal, S., Weatherall, D. J. and Kulozik, A. E. (1989). The frequency and origin of the sickle cell mutation in the district of Coruche/Portugal. *Human Genetics*, **82**, 255–8.

Nakamura, Y., Leppert, M., O'Connell, P., Wolff, R., Holm, T., Culver, M., Martin, C., Fujimoto, E., Hoff, M., Kumlin, E. and White, R. (1987). Variable number of tandem repeat (VNTR) markers for human gene mapping. *Science*, **235**, 1616–22.

Nakatsuji, T., Landman, H. and Huisman, T. H. (1986). An elongated segment of DNA observed between two human alpha globin genes. *Human Genetics*, **74**, 368–71.

Neel, J. V. (1978). The population structure of an Amerindian tribe, the Yanomama. *Annual Review of Genetics*, **12**, 365–413.

Nei, M. and Roychoudhury, A. K. (1982). Genetic relationships and evolution of human races. *Evolutionary Biology*, **14**, 1–59.

Noronha, P. A. and Honig, G. R. (1978). β thalassaemia arising as a new mutation in an American child. *American Journal of Hematology*, **4**, 187–92.

Nurse, G. T. (1985). The pace of human selective adaptation to malaria. *Journal of Human Evolution*, **14**, 319–26.

O'Shaughnessy, D. R. (1989). *Haemoglobin Variants in Oceania*. D. Phil. Thesis, University of Oxford.

O'Shaughnessy, D. R., Hill, A. V. S., Bowden, D. K., Weatherall, D. J. and Clegg, J. B. (1990). Globin gene in Micronesia: origin and affinities of Pacific Island peoples. *American Journal of Human Genetics*, **46**, 144–55.

Oehme, R., Kohne, E. and Horst, J. (1985). DNA polymorphic patterns linked to β globin genes in German families affected with haemoglobinopathies and β thalassaemias: a comparison to other ethnic groups. *Human Genetics*, **71**, 219–22.

Old, J. M., Heath, C., Fitches, A., Thein, S. L., Jeffreys, A. J., Petrou, M., Modell, B. and Weatherall, D. J. (1986). Meiotic recombination between two polymorphic restriction sites within the β globin gene cluster. *Journal of Medical Genetics*, **23**, 14–18.

Old, J. M., Petrou, M., Modell, B. and Weatherall, D. J. (1984). Feasibility of antenatal diagnosis of β thalassaemia by RNA polymorphisms in Asian Indian and Cypriot populations. *British Journal of Haematology*, **57**, 255–63.

Orkin, S. H. and Goff, S. C. (1981). Nonsense and frameshift mutations in β thalassaemia detected in cloned β globin genes. *Journal of Biological Chemistry*, **256**, 9782–4.

Orkin, S. H. and Kazazian, H. H. (1984). The mutation and polymorphism of the human β-globin gene and its surrounding DNA. *Annual Review of Genetics*, **18**, 131–71.

Orkin, S. H., Kazazian, H. H., Antonarakis, S. E., Goff, S. C., Boehm, C. D., Sexton, J. P., Waber, P. G. and Giardina, P. J. V. (1982). Linkage of β thalassaemia mutations and β globin polymorphisms with DNA polymorphisms in human β globin gene cluster. *Nature*, **296**, 627–31.

Orr-Weaver, T. L. and Szostak, J. W. (1985). Fungal recombination. *Microbiology Reviews*, **49**, 33–58.

Ottolenghi, S. and Carestia, C. (1986). β globin gene disorders in Italy and the Mediterranean area. In *Human Genes and Diseases*, ed. F. Blasi, pp. 257–98. New York: John Wiley and Sons.

Pagnier, J., Dunda-Belkhodja, O., Zohoun, I., Teyssier, J., Baya, H., Jaeger, G., Nagel, R. L., Labie, D. (1984b). α thalassaemia among sickle cell anaemia patients in various African populations. *Human Genetics*, **68**, 318–319.

Pagnier, J., Mears, J. G., Dunda-Belkhodja, O., Schaefer-Rego, K. E., Beldjord, C., Nagel, R. L. and Labie, D. (1984a). Evidence for the multicentric origin of the sickle cell hemoglobin gene in Africa. *Proceedings of the National Academy of Sciences USA*, **81**, 1771–3.

Pawley, A. and Green, R. C. (1984). The proto-Oceanic language community. *Journal of Pacific History*, **19**, 123–46.

Pirastu, M., Galanello, R., Doherty, M. A., Tuveri, T., Cao, A. and Kan, Y. W. (1987). The same β globin gene mutation is present on nine different β thalassaemia chromosomes in a Sardinian population. *Proceedings of the National Academy of Sciences USA*, **84**, 2882–5.

Pirastu, M., Saglio, G., Camaschella, C., Loi, A., Serra, A., Bertero, T., Gabutti, W. and Cao, A. (1988). Delineation of specific β thalassaemia mutations in high risk areas of Italy: a pre-requisite for prenatal diagnosis. *Blood*, **71**, 983–8.

Plato, C. C., Rucknagel, D. L. and Gershowitz, H. (1964). Studies on the distribution of glucose–6-phosphate dehydrogenase deficiency, Thalassemia and other genetic traits in coastal and mountain villages of Cyprus. *American Journal of Human Genetics*, **27**, 198–212.

Poncz, M., Schwartz, E., Ballantine, M. and Surrey, S. (1983). Nucleotide sequence analysis of the delta-beta globin gene region in humans. *Journal of Biological Chemistry*, **258**, 11599–609.

Powers, P. A., Altay, C., Huisman, T. H. J. and Smithies, O. (1984). Two novel arrangements of the human fetal globin gene $^Gγ$-$^Gγ$ and $^Aγ$-$^Aγ$. *Nucleic Acids Research*, **12**, 7023–34.

Proudfoot, N. J., Gil, A. and Maniatis, T. (1982). The structure of the human ζ-globin gene and a closely linked nearly identical pseudogene. *Cell*, **31**, 533–63.

Ramsay, M. and Jenkins, T. (1987). Globin gene associated restriction fragment length polymorphisms in Southern African Peoples. *American Journal of Human Genetics*, **41**, 1132–44.

Ramsay, M. and Jenkins, T. (1988). Alpha-globin gene cluster haplotypes in the Kalahari San and Southern African Bantu speaking blacks. *American Journal of Human Genetics*, **43**, 527–33.

Raper, H. B. (1956). Sickling in relation to morbidity from malaria and other diseases. *British Medical Journal*, **1**, 965–6.

Rosatelli, C., Falchi, A. M., Tuveri, T., Scalas, M. T., DiTucci, A., Monni, G. and Cao, A. (1985). Prenatal diagnosis of β thalassaemia with synthetic oligomer technique. *Lancet*, **i**, 241–3.

Rosatelli, C., Leoni, G. B., Tuveri, T., Scalas, M. T., DiTucci, A. and Cao, A. (1987). β thalassaemia in Sardinians: implications for prenatal diagnosis. *Journal of Medical Genetics*, **24**, 97–100.

Rosatelli, C., Tuveri, T., Scalas, M. T., DiTucci, A., Leoni, G. B., Furbetta, A., Monni, A. and Cao, A. (1988). Prenatal diagnosis of β thalassaemia by oligonucleotide analysis in Mediterranean populations. *Journal of Medical Genetics*, **25**, 762–5.

Rouabhi, L., Lapoumeroulie, C., Amselem, S., Krishnamoorthy, R., Adjrad, L., Girot, R., Chardin, P., Benabdji, M., Labie, D. and Beldjord, C. (1988). DNA haplotype distribution in Algerian β thalassaemia patients. An extended evaluation by family studies and representative molecular characterization. *Human Genetics*, **79**, 373–6.

Rucknagel, D. L. and Neel, J. V. (1961). The haemoglobinopathies. *Progress in Medical Genetics*, **1**, 158–260.

Sampeitro, M., Cappellini, M. D., Fiorelli, G., Wainscoat, J. S., Thein S. L. and Weatherall, D. J. (1988). Genotypes of thalassaemia major and intermedia in Italy. *Birth Defects*, **23**, 117–23.

Savatier, P., Trabuchet, G., Chebloune, Y., Faure, C., Verdier, G. and Nigon, V. M. (1987). Nucleotide sequence of the beta globin genes in Gorilla and Macaque: the origin or nucleotide polymorphisms in human. *Journal of Molecular Evolution*, **24**, 309–18.

Semenza, G. L., Malladi, P., Poncz, M., Degrosso, K., Schwartz, E. and Surray, S. (1984). Detection of a novel DNA polymorphism in the β globin cluster and evidence from site specific recombination. *Clinical Research*, **18**, 225A.

Shimasaki, S. and Suchi, I. (1986). Diversity of human α globin gene loci including quadruplicated arrangement. *Blood*, **67**, 784–8.

Shimizu, K. (1987). Characteristics of β^A chromosome haplotypes in Japanese. *Biochemical Genetics*, **25**, 197–203.

Simmons, R. T. and Booth, P. D. (1971). *A compendium of Melanesian genetic data I–IV*. Victoria: Commonwealth Serum Laboratories.

Siniscalco, M., Bernini, L., Latte, B. and Motulsky, A. G. (1961). Favism and Thalassaemia in Sardinia and their relationship to malaria. *Nature*, **190**, 1179–80.

Siniscalco, M., Bernini, L., Filippi, G., Latte, B., Khan, P. M., Piomelli, S. and Rattazzi, M. (1966). Population genetics of haemoglobin variants, thalassaemia and glucose-6-phosphate dehydrogenase deficiency, with particular reference to the malaria hypothesis. *Bulletin of the World Health Organization*, **34**, 379–93.

Slightom, J. L., Blechl, A. E. and Smithes, O. (1980). Human fetal $^G\gamma$ and $^A\gamma$ globin genes: complete nucleotide sequences suggest that DNA can be exchanged between duplicated genes. *Cell*, **21**, 627–38.

Slightom, J. L., Koop, B. F., Xu, P. and Goodman M. C. (1988). Rhesus fetal globin genes. Concerted gene evolution in the descent of higher primates. *Journal of Biological Chemistry*, **263**, 12427–38.

Smith, F. H. and Spencer, F. (ed.) (1984). *The Origin of Modern Humans: a World Survey of the Fossil Evidence*. New York: Alan R. Liss.

Smithies, O. and Powers, P. A. (1986). Gene conversions and their relation to homologous chromosome pairing. *Philosophical Transactions of the Royal Society of London B*, **312**, 291–302.

Solomon, E. and Bodmer, W. F. (1979). Evolution of sickle variant gene. *Lancet*, i, 923.

Sozuoz, A., Berkalp, A., Figus, A., Loi, A., Pirastu, M. and Cao, A. (1988). β thalassaemia mutations in Turkish Cypriots. *Journal of Medical Genetics*, **25**, 766–8.

Spriggs, M. (1984). The Lapita cultural complex. *Journal of Pacific History*, **19**, 185–206.

Spritz, R. A. (1981). Duplication deletion polymorphisms 5' to the human β globin gene. *Nucleic Acids Research*, **9**, 5037–47.

Stamatoyannopoulos, G. and Fessas, P. (1964). Thalassaemia glucose-6-phosphate dehydrogenase deficiency, sickling and malaria endemicity in Greece; a study of five areas. *British Medical Journal*, **1**, 875–9.

Stamatoyannopoulos, G. and Nienhuis, A. W. (ed.) (1987). *Developmental Control of Globin Gene Expression*. New York: Alan R. Liss.

Stamatoyannopoulos, G., Nienhuis, A. W., Leder, P. and Majerus, P. W. (1987). *The Molecular Basis of Blood Diseases*. Philadelphia: W. B. Saunders.

Stamatoyannopoulos, G., Nute, P. E. and Miller, M. (1981). De novo mutations producing unstable haemoglobin or haemoglobins. *Human Genetics*, **58**, 396–9.

Starck, J., Bouhass, R., Morle, F. and Godet, J. (1990). Extent and high frequency of a short conversion between the human $^A\gamma$ and $^G\gamma$ fetal globin genes. *Human Genetics*, **84**, 179–84.

Stringer, C. B. and Andrews, P. (1988). Genetic and fossil evidence for the origin of modern humans. *Science*, **239**, 1263–8.

Sukumaran, P. K., Nakatsuji, T., Gardiner, M. B., Reese, A. L., Gilman, J. G. and Huisman, T. H. J. (1983). Gamma thalassaemia resulting form the deletion of a $^G\gamma$ globin gene. *Nucleic Acids Research*, **11**, 4635–43.

Tautz, D. (1989). Hypervariability of simple sequences as a general source for polymorphic DNA markers. *Nucleic Acids Research*, **17**, 6463–71.

Thein, S. L. Hesketh, C., Wallace, R. B. and Weatherall, D. J. (1988). The molecular basis of thalassaemia major and thalassaemia intermedia in Asian Indians: application to prenatal diagnosis. *British Journal of Haematology*, **70**, 225–31.

Thein, S. L., Old, J. M., Wainscoat, H. S. Petrou, M., Modell, B. and Weatherall, D. J. (1984). Population and genetic studies suggest a single origin for the Indian deletion β°-thalassaemia. *British Journal of Haematology*, **57**, 271–8.

Tonz, O., Glatthaar, B. E., Winterhalter, K. H. and Ritter, H. (1973). New mutation in a Swiss girl leading to a clinical and biochemical β thalassaemia minor. *Human Genetics*, **20**, 321–7.

Trecartin, R. F., Liebhaber, S. A., Chang, J. C., Lee, Y. W. and Kan, Y. W. (1981). β°-thalassaemia in Sardinia is caused by a nonsense mutation. *Journal of Clinical Investigation*, **68**, 1012–17.

Treco, D., Thomas, B. and Arnheim, N. (1985). Recombination hot spot in the human β globin gene cluster: meiotic recombination of human DNA fragments in *Saccharomyces cerevisiae*. *Molecular and Cellular Biology*, **5**, 2029–38.

Trent, R. J., Bowden, D. K., Old, J. M., Wainscoat, J. S., Clegg, J. B. and Weatherall, D. J. (1981a). A novel rearrangement of the human beta-like globin gene cluster. *Nucleic Acids Research*, **9**, 6723–33.

Trent, R. J., Buchanan, J. B., Webb, A., Goundar, R. P. S., Seruvatu, L. M. and Mickleson, K. N. P. (1988). Globin genes are useful markers to identify genetic similarities between Fijians and Pacific Islanders from Polynesia and Melanesia. *American Journal of Human Genetics*, **42**, 601–7.

Trent, R. J., Higgs, D. R., Clegg, J. B. and Weatherall, D. J. (1981b). A new triplicated α globin gene arrangement in man. *British Journal of Haematology*, **49**, 149–52.

Trent, R. J., Mickleson, K. N. P., Wilkinson, T., Yakas, J., Bluck, R., Dixon, M., Liley, A. W. and Kronenberg, H. (1985). α globin gene rearrangements in Polynesians are not associated with malaria. *American Journal of Hematology*, **18**, 431–3.

Trent, R. J., Mickleson, K. N. P., Wilkinson, T., Yakas, J., Dixon, M. W., Hill, P. J. and Kronenberg, H. (1986). Globin genes in Polynesians have many rearrangements including a recently described $\gamma\gamma\gamma\gamma$. *American Journal of Human Genetics*, **39**, 350–60.

Tsintsof, A. S., Hertzberg, M. S., Prior, J. F., Mickleson, K. N. P. and Trent, R. J. (1990). α globin gene markers identify genetic differences between Australian Aborigines and Melanesians. *American Journal of Human Genetics*, **46**, 138–43.

Van Valen, L. M. (1986). Speciation and our own species. *Nature*, **322**, 412.

Wainscoat, J. S., Bell, J. I., Old, J. M., Weatherall, D. J., Furbetta, M., Galanello, R. and Cao, A. (1983a). Globin gene mapping studies in Sardinian patients homozygous for β°-thalassaemia. *Molecular and Biological Medicine*, **1**, 1–10.

Wainscoat, J. S., Bell, J. I., Thein, S. L., Higgs, D. R., Serjeant, G. R., Peto, T. E. A. and Weatherall, D. J. (1983b). Multiple origins of the sickle mutation: evidence from β^{S}-globin gene cluster polymorphisms. *Molecular Biology and Medicine*, **1**, 191–7.

Wainscoat, J. S., Hill, A. V. S., Boyce, A. J., Flint, J., Hernandez, M., Thein, S. L., Old, J. M., Lynch, J. R., Falusi, A. G., Weatherall, D. J. and Clegg, J. B. (1986a). Evolutionary relationships of human populations from an analysis of nuclear DNA polymorphisms. *Nature*, **319**, 491–3.

Wainscoat, J. S., Hill, A. V. S., Thein, S. L., Weatherall, D. J., Clegg, J. B. and Higgs, D. R. (1989). Geographic distribution of α and β globin gene cluster polymorphisms. In *The Origins and Dispersal of Modern Humans: Behavioural and Biological Perspectives*, ed. P. Mellars, and C. B. Stringer, pp. 31–8. Edinburgh: Edinburgh University Press.

Wainscoat, J. S., Old, J. M., Weatherall, D. J. and Orkin, S. H. (1983c) The molecular basis for the clinical diversity of β thalassaemia in Cypriots. *Lancet*, **i**, 1235–7.

Wainscoat, J. S., Work, S., Sampietro, M., Cappellini, M. D., Fiorelli, G., Terzoli, S. and Weatherall, D. J. (1986b). Feasibility of prenatal diagnosis of β thalassaemia by DNA polymorphisms in an Italian population. *British Journal of Haematology*, **62**, 495–500.

Weatherall, D. J. (1986). The Thalassaemias: molecular pathogenesis. In *Hemoglobin: Molecular, Genetic and Clinical Aspects*, ed. H. T. Bunn and B. G. Forget, pp. 213–321. Philadelphia: W. B. Saunders.

Weber, J. L. and May, P. E. (1989). Abundant class of human DNA polymorphism which can be typed using the polymerase chain reaction. *American Journal of Human Genetics*, **44**, 388–96.

Weir, B. S. and Hill, W. G. (1986). Nonuniform recombination within the human β globin gene cluster. *American Journal of Human Genetics*, **38**, 776–8.

White, P. J. and O'Connell, J. F. (1982). *A prehistory of Australia, New Guinea and Sahul*. Sydney: Academic Press.

Winichagoon, P., Higgs, D. R., Goodbourn, S. E. Y., Lamb, J., Clegg, J. B. and Weatherall, D. J. (1982). Multiple rearrangements of the human embryonic zeta globin genes. *Nucleic Acids Research*, **10**, 5853–68.

Wolpoff, M. H. (1989). Multi-regional evolution: the fossil alternative to Eden. In *The Origins and Dispersal of Modern Humans: Behavioural and Biological Perspectives*, ed. P. Mellars and C. B. Stringer, pp. 62–108. Edinburgh: Edinburgh University Press.

Wong, C., Antonarakis, S. E., Goff, S. C., Orkin, S. H., Boehm, C. D. and Kazazian, H. H. (1986). On the origin and spread of β thalassaemia: recurrent observation of four mutations in different ethnic groups. *Proceedings of the National Academy of Sciences USA*, **83**, 6529–32.

Woo, S. L. C. (1988). Collation of RFLP haplotypes at the human phenylalanine hydroxylase (PAH) locus. *American Journal of Human Genetics*, **43**, 781–3.

Wurm, S. A. (1982). *Papuan Languages of Oceania*. Tübingen: Gunther Narr.
(1983). Linguistic prehistory in the New Guinea area. *Journal of Human Evolution*, **12**, 25–35.

Yenchitsomanus, P., Summers, K. M., Bhatia, K. K. and Board, P. G. (1986b). A single α globin gene deletion in Australian Aborigines. *Australian Journal of Experimental Biology and Medical Science*, **64**, 297–306.

Yenchitsomanus, P., Summers, K. M., Board, P. G., Bhatia, K. K., Jones, G. L., Johnston, K. and Nurse, G. T. (1986a). Alpha thalassaemia in Papua New Guinea. *Human Genetics*, **74**, 432–7.

Yenchitsomanus, P., Summers, K. M., Board, P. G., Fucharoen, S. and Wasi, P. (1988). DNA polymorphisms of β^N and β^E globin genes in Thais. *Birth Defects*, **23**, 99–106.

Yongvanit, P., Sriboonlue, P., Mularlee, N., Karnthong, T., Areetjitranusorn, P., Hundrieser, J., Limberg, R., Schulze, B., Laig, M., Flatz, S. D. and Flatz, G. (1989). DNA haplotypes and frameworks linked to the β globin locus in an Astray–Asiatic population with a high prevalence of hemoglobin E. *Human Genetics*, **83**, 171–4.

Zeng, Y.-T. and Huang, S.-Z. (1987). Disorders of haemoglobin in China. *Journal of Medical Genetics*, **24**, 578–83.

Zhang, J.-Z., Cai, S. P., He, X., Lin, H.-X., Lin, H.-J., Huang, Z.-G., Chehab, F. F. and Kan, Y. W. (1988). Molecular basis of β thalassaemia in South China: strategy for DNA analysis. *Human Genetics*, **78**, 37–40.

6 Mitochondrial DNA: its uses in anthropological research

DON J. MELNICK, GUY A. HOELZER AND
RODNEY L. HONEYCUTT

Introduction

With the technological revolution in molecular biology of the past 15
years that allows a broader and more detailed examination of nucleic acid
sequences than many thought would ever be possible (Nei, 1987; Hillis
and Moritz, 1990), recombinant DNA methods have made accessible
large portions of the nuclear and organelle genomes of a vast array of
plant and animal species. A large number of evolutionary studies, for
reasons outlined in the next section, have focused on the DNA found in
the mitochondria (mtDNA) of the cells of eukaryotes. These studies
concern the taxonomic relationships between humans and their closest
relatives, the African great apes (e.g. Brown *et al.*, 1982), and attempt to
identify the age and geographical location of the common ancestor of all
modern humans (e.g. Brown, 1980; Johnson *et al.*, 1983; Cann *et al.*,
1987), have received a great deal of attention (Gibbons, 1990).

This chapter introduces the methods and some applications of mtDNA
analysis. There are excellent reviews of the molecular biology of mtDNA
(Attardi, 1985; Brown, 1983, 1985; Clayton, 1982, 1984) and two comprehensive surveys of the research on human and non-human primate
mtDNA (Spuhler, 1988; Honeycutt and Wheeler, 1989). This chapter
should be viewed as an updated, non-technical description of the same
material, with an emphasis on primate (including human) population and
evolutionary genetics. It will describe (1) the mitochondrial genome, (2)
methods used to detect and quantify variation in mtDNA, and (3)
applications of mtDNA analysis to evolutionary problems of anthropological interest, including the controversies surrounding hominoid and
modern human phylogenetic reconstructions.

The mitochondrial genome

General description

Mitochondria are small bodies or organelles in the cytoplasm of a
eukaryotic cell. Their primary function is to provide cellular energy

179

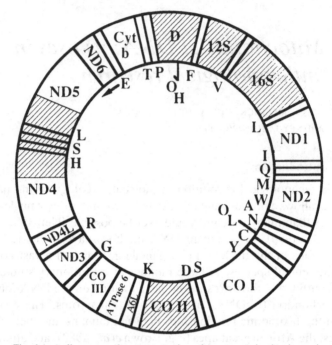

Fig. 6.1. A diagrammatic representation of gene organization in a primate mtDNA molecule. The origins of replication for the heavy (OH) and light (OL) strands are denoted by hatch marks on the inside of the circle. Upper case letters along the inside rim of the circle indicate the locations of tRNA genes. The positions of other genes are listed in the ring. Hatched regions indicate those portions of the mtDNA genome for which comparative sequence data exist for primates.

through respiration and oxidation. Unlike most other organelles (chloroplasts being the other exception), mitochondria contain their own haploid genomes, leading to speculation that they are descended from a symbiotic bacterium that found its way into an early ancestor of the eukaryotic cell (Margulis, 1981). Much of what is known about the organization of the mitochondrial genome comes from research on a few vertebrate species, where the nucleotide composition and order of the entire mtDNA molecule has been completely described.

The mtDNA of mammals (Fig. 6.1), including humans and other primates, is a small, duplex (double-stranded), closed-circular DNA molecule, with an average size of approximately 16 400 base pairs (Brown, 1983, 1985; Honeycutt and Wheeler, 1989). In size, organization and nucleotide sequence composition, the mammalian mitochondrial genome depicts simplicity in design and efficiency in sequence use. Thus, unlike nuclear DNA, mtDNA consists primarily of coding sequences (*c.* 90%), but lacks split genes (i.e. there are no introns), repetitive sequences (except in the non-coding D-loop region), and many intergenic sequences (Anderson *et al.*, 1981, 1982; Bibb *et al.*, 1981; Clary *et al.*, 1982; Roe *et al.*, 1985; Wolstenholme and Clary, 1985). The constitution of human mitochondrial DNA is summarized by Wallace (1989).

Both gene content and order within the vertebrate mitochondrial genome are highly conserved, with the major coding regions consisting of:

1. *Two* ribosomal RNA genes (12S and 16S rRNAs);
2. *22* transfer RNAs (tRNAs) required for mitochondrial protein synthesis; and
3. *13* structural protein genes (cytochrome *c* oxidase subunits I, II and III, ATPase subunit 6, cytochrome *b*, seven genes encoding subunits of the NADH dehydrogenase complex (ND-1 to 4, 4L, 5, 6), and one gene, A6L, coding for a subunit of the ATP synthetase complex) critical to mitochondrial respiration (Anderson *et al.*, 1981; Chomyn *et al.*, 1985, 1986).

In addition, there is a major non-coding region, the D-loop, located between the tRNA^{Pro} and the tRNA^{Phe} genes (Anderson *et al.*, 1981, 1982; Bibb *et al.*, 1981). This particular region is involved in both transcription and replication of the mtDNA molecule (Clayton, 1982, 1984; Attardi, 1985; Brown, 1985). Because the circular arrangement of all of these sequences (Fig. 6.1) is shared between amphibians and mammals (Brown, 1985), it must date back at least to the split between anamniote and amniote tetrapods some 340–350 million years ago (Olson, personal communication).

Besides its economy of organization and small size, the mitochondrial genome has several other unique features that make it an ideal molecule for use in evolutionary study. First, the mitochondrial genetic code of vertebrates differs in codon usage from the 'universal' code seen in the nuclear genome (Barrell *et al.*, 1979; Anderson *et al.*, 1981). Second, mtDNA is maternally inherited in a clonal fashion through the egg cytoplasm (Dawid and Blackler, 1972; Hutchison *et al.*, 1974; Giles *et al.*, 1980; Reilly and Thomas, 1980; Case and Wallace, 1981; Lansman *et al.*, 1983; Gyllensten *et al.*, 1985). Third, there is no strong evidence of inter-molecular recombination (DeFrancesco *et al.*, 1980; Solus and Eisen-stadt, 1984; Hayashi *et al.*, 1985), though there is some evidence of rearrangements in gene order (Brown, 1985; Moritz *et al.*, 1987; Yang and Zhou, 1988), duplication of both non-coding and coding regions

182   D. J. Melnick, G. A. Hoelzer and R. L. Honeycutt

within a given taxon (Moritz and Brown, 1986, Moritz, 1987), and small deletions/additions within particular regions of the genome (Cann and Wilson, 1983). Both clonal inheritance of a maternally transmitted haploid genome and a lack of recombination greatly simplify the determination of phylogenetic relationships and the determination of such population parameters as migration rates (see below). Fourth, despite the numerous copies of mtDNA within an individual, as a rule they are all identical (a condition known as homoplasmy or the presence of one haplotype). Some length and site heteroplasmy (variation in size or sequence within an individual) has been observed in several organisms (Solignac *et al.*, 1983; Densmore *et al.*, 1985; Bermingham *et al.*, 1986; Hale and Singh, 1986; Rand and Harrison, 1986; Bentzen *et al.*, 1988) but, for the most part, they are rare in mammals (for exceptions see Hauswirth *et al.*, 1984; Boursot *et al.*, 1987; Honeycutt *et al.*, 1991). Finally, and somewhat paradoxically, despite a highly conservative gene arrangement, the mitochondrial genome of primates and other mammals has experienced a rate of nucleotide base substitution (or molecular evolution) some five to ten times that found in the nuclear genome (Brown *et al.*, 1979). Thus, for any specific study, a detailed analysis of the mitochondrial genome is likely to yield a greater amount of genetic variation, with which to answer evolutionary questions, and which is particularly useful when studying members of the same or closely related taxa.

Primate mtDNA

Seven complete vertebrate mitochondrial genomes have been sequenced: those of five mammals, one bird, and one amphibian, as follows: (1) human (*Homo sapiens*: Anderson *et al.*, 1981); (2) house mouse (*Mus domesticus*: Bibb *et al.*, 1981); (3) Norway rat (*Rattus norvegicus*: Gadaleta *et al.*, 1989); (4) cow (*Bos taurus*: Anderson *et al.*, 1982); (5) Fin whale (*Balaenoptera physalus*: Arnason *et al.*, 1991; (6) African clawed-frog (*Xenopus laevis*: Roe *et al.*, 1985); (7) domestic chicken (*Gallus domesticus*: Desjardins and Morais, 1990). These sequence data are valuable, because they provide a broad picture of conserved and variable features of vertebrate mtDNA. However, from a phylogenetic standpoint, sequences from these divergent taxa cannot provide detailed information on the rates of mtDNA evolution among vertebrates, and provide only a limited amount of information on the distribution and frequency of nucleotide substitutions within the class Mammalia.

By far the most extensive comparative data on mtDNA sequence variation among coding and non-coding regions exist for the orders Primates and Artiodactyla (Table 6.1), though even in these two orders the amount of sequence data and number of taxa are limited. Therefore, there is little information on mtDNA sequence variation that can be used to evaluate the effectiveness of a particular mtDNA region for resolving phylogenetic branching patterns in mammals. This lack of comparative data does not mean that mtDNA sequence information is of no value to mammalian systematics. On the contrary, the differential rates of nucleotide substitutions involving transitions and transversions, and the rate of amino acid replacements in protein coding regions suggest that mtDNA sequence variation can be used over a broad range of phylogenetic divergence (Wilson *et al.*, 1985; Miyamoto and Boyle, 1989), so that it can be used to examine variation at both higher and lower levels of divergence.

The most taxonomically diverse, detailed comparative studies using nucleotide sequences have been conducted on an 896 base pair (bp) fragment containing the ND-4 and ND-5 genes and three tRNA genes in 13 primate species (Table 6.1; Brown *et al.*, 1982; Hayasaka *et al.*, 1988a). These data allow for a comparison of homologous coding regions in a monophyletic group that spans a broad range of evolutionary time. The results from these comparisons are: (1) There is a predominance of single base substitutions relative to base deletions/additions. (2) Silent substitutions are four to six times more frequent than amino acid replacements. (3) The frequency of substitutions at codon positions is highest at the third position and lowest at the second position. (4) There is an initial nucleotide substitution bias toward transitions (A----G; C----T) over transversions (A or G-----C or T), with transitions occurring at a frequency of 92–95% among recently diverged taxa. (5) Transition to transversion ratios decreases with an increase in divergence time (greater than 30 million years), possibly as a result of an increase in multiple substitutions at a nucleotide site. (6) Sequence similarity ranges from 70–97% between species pairs. (7) The resultant phylogenetic tree constructed from the overall sequence data is congruent with results from other phylogenetic studies using molecules and morphology. Other studies on primates (Aquadro and Greenberg, 1983; Hixson and Brown, 1986; Ruvolo *et al.*, 1991; Disotell *et al.*, 1992), artiodactyls (Tanhauser, 1985; Miyamoto and Boyle, 1989), and rodents (Allard, 1990; Allard and Honeycutt, 1992) confirm most of the patterns observed in the 896 bp fragment. The only exception is that the transition/transversion ratios are

Table 6.1. *Coding and non-coding regions of mammalian mtDNA that have been sequenced to date*

Taxa	Region	Reference
Primates		
Homo sapiens	Entire genome	Anderson *et al.*, 1981
Homo sapiens	200 bp COIII	Monnat and Loeb, 1985
Homo sapiens	Intergenic region	Wrischnik *et al.*, 1987
Homo, Gorilla, Pan	D-loop	Aquadro and Greenberg, 1983
		Foran *et al.*, 1988
Homo, Gorilla, Pan	16S rRNA	Hixson, unpublished data
Homo, Gorilla, Pan, Pongo,	12S rRNA	Hixson and Brown, 1986
Macaca		D. J. Melnick *et al.*, unpublished
		data
Homo, Gorilla, Pan, Hylobates,	COII	Anderson *et al.*, 1981
Macaca, Papio,		Ramharack and Deeley, 1987
Cercophithecus		Ruvolo *et al.*, 1991
Theropithecus, Mandrillus,	COII	Disotell *et al.*, 1992
Cercocebus		
Hominoid primates, *Macaca,*	ND4 & 5, 3 tRNAs	Brown *et al.*, 1982
Saimiri, Tarsius, Lemur		Hayasaka *et al.*, 1988a
Chiroptera		
Pteropus	COIII	Bennet *et al.*, 1988
Rousettus, Phyllostomus	COII	Adkins and Honeycutt, 1991
Dermoptera		
Cynocephalus	COII	Adkins and Honeycutt, 1991
Scandentia		
Tupaia	COII	Adkins and Honeycutt, 1991
Cetacea		
Balaenoptera physalus	Entire genome	Arnason *et al.*, 1991
Cephalorhynchus	Cytochrome b,	Southern *et al.*, 1988
	ND3	
Stenella	Cytochrome b	Irwin *et al.*, 1991
Artiodactyla		
Bos taurus	Entire genome	Anderson *et al.*, 1982
Bison, Bubalus, Bos, Giraffa,	12S rRNA, 16S	Tanhauser, 1985
Sus, Cervis, Muntiacus,	rRNA, 3 tRNAs	Miyamoto *et al.*, 1989, 1990
Odocoileus, Hydropotes		
Ovis, Antilocapra, Giraffa,	Cytochrome b	Irwin *et al.*, 1991
Dama, Odocoileus, Tragulus,		
Camelus, Sus, Capra		
Perissodactyla		
Equus, Diceros	Cytochrome b	Irwin *et al.*, 1991
Quagga	NDI, COI	Higuchi *et al.*, 1984
Proboscidea		
Loxodonta	Cytochrome b	Irwin *et al.*, 1991

Table 6.1. (*cont.*)

Taxa	Region	Reference
Edentata		
Dasypus	COII	Adkins and Honeycutt, unpublished data
Rodentia		
Mus domesticus, *Rattus* norvegicus	Entire genome	Bibb *et al.*, 1981 Gadaleta *et al.*, 1989
Dipodomys	D-loop, 2 tRNAs	Thomas *et al.*, 1990
Georychus, *Cryptomys*, *Bathyergus*, *Heterocephalus*, *Heliophobius*, *Thryonomys*, *Petromus*, *Hystrix*, *Myocastor*, *Chinchilla*, *Octodon*, *Cavia*, *Capromys*	12S rRNA	Allard and Honeycutt, 1992
Rattus	12S rRNA, 2 tRNAs	Kobayashi *et al.*, 1981
Rattus	13 tRNAs	Cantatore *et al.*, 1982
Rattus	Cytochrome b, 3 tRNAs	Koike *et al.*, 1981 Gortz and Feldman, 1982
Rattus	4 tRNAs	Wolstenholme *et al.*, 1982
Rattus	COI, II, III, ATPase 6, tRNAs	Grosskopf and Feldman, 1981
Rattus	COII	Brown and Simpson, 1982

lower for more closely related non-primates than primates (Allard, 1990).

Most of the information on nucleotide substitution rates in mammalian mtDNA is also derived from primates (Brown *et al.*, 1979, 1982; Ferris *et al.*, 1981). Comparison of overall restriction endonuclease site variation in selected primate taxa revealed that there was five to ten times as much divergence in mtDNA as in single-copy nuclear DNA (scnDNA). When indirect estimates of nucleotide sequence divergence (d) were compared with known divergence times for particular pairs of primates, the initial rate was estimated to be approximately 0.02 substitutions per base pair per million years. Subsequent annealing experiments and nucleotide sequencing have confirmed the rate estimates from restriction site mapping.

Coding and non-coding regions of the mitochondrial genome also evolve at different rates (Brown *et al.*, 1982; Ferris *et al.*, 1983; Cann *et al.*, 1984; Brown, 1985). For instance, mitochondrial tRNA and rRNA genes, which evolve faster than their nuclear counterparts, have very low rates of nucleotide substitution relative to the non-coding region of the mitochondrial genome (Brown *et al.*, 1982; Cann *et al.*, 1984; Hixson and Brown, 1986). Most mitochondrial protein genes are intermediate in rates of change, with the NADH dehydrogenase genes evolving faster than other genes such as the cytochrome c oxidase subunit genes. By far the most divergent parts of the mitochondrial genome are the non-coding regions such as the D-loop. Rates of D-loop divergence are 1.4 to 5 times those of protein genes (Aquadro and Greenberg, 1983; Brown, 1985; Foran *et al.*, 1988; Vigilant *et al.*, 1989).

Conclusion. The small size, conserved organization, pattern of sequence evolution, mode of inheritance, and combination of rapidly and slowly evolving regions are all properties that make mtDNA an important molecule for detailed evolutionary studies both within and between species of primates, as well as other mammals. These same characteristics also make it relatively easy to manipulate the mtDNA genome in the laboratory and to analyse quantitatively and interpret empirical data. All of these factors are undoubtedly the reason for the plethora of evolutionary studies over the past 5–10 years that have used mtDNA as their source of information.

Mitochondrial DNA analysis

Patterns of mtDNA variation have been examined at several levels of divergence in primates ranging from studies of local populations to comparisons of genera and families. While the details of any particular mtDNA study on primates may vary depending upon the questions asked, all experimental approaches are concerned with an assessment of nucleotide sequence variation. This section briefly describes the methods used to isolate mtDNA, uncover differences between individuals and analyse quantitatively the observed variation.

Isolation of mtDNA

Mitochondrial DNA can be isolated from a variety of tissues including heart, liver, kidneys, blood, skin biopsies, hair, and even museum specimens. Some analyses require a highly purified form of mtDNA, whereas others need only cruder preparations of 'total DNA'. The advent of the polymerase chain reaction (PCR) has made many comparative studies more feasible and much easier to conduct, and in the future the more laborious approaches to mtDNA isolation and analysis may become obsolete.

Isolation of highly purified mtDNA with buoyant density centrifugation. Many studies of mtDNA variation in primates have involved restriction enzyme digestions of highly purified mtDNA (e.g. Brown, 1981; Ferris *et al.*, 1981a,b; Stoneking *et al.*, 1986; Cann *et al.*, 1987). This mtDNA is isolated using buoyant density centrifugation over a continuous caesium chloride (CsCl) gradient (Brown, 1981; Lansman *et al.*, 1981; Densmore *et al.*, 1985; Dowling *et al.*, 1990). The procedure is as follows. (1) Tissues are homogenized and intact mitochondria are isolated from other cellular components, including most of the nuclear DNA. (2) The mitochondria are lysed and placed in a CsCl gradient with an intercalating dye (either ethidium bromide or propidium iodide). (3) After ultracentrifugation at 36 000 rpm for approximately 24 hours, nuclear and mitochondrial DNA form bands in the gradient according to their respective densities. Because circular mtDNA molecules have a higher density than linear nuclear DNA, the mtDNA band maintains a position below the nuclear DNA band. (4) Under ultraviolet wavelengths the intercalating dye appears fluorescent, making both DNA bands visible. The mtDNA band can be recovered and re-banded in a discontinuous CsCl step-gradient at 45 000 rpm. (5) Following a second continuous CsCl gradient, the salts and intercalating dye are removed by either extraction with isopropanol and dialysis, or filtration of high molecular weight DNA in centrifugation columns.

Although a very pure form of mtDNA can be obtained by this method of isolation, the procedure is extremely time-consuming and impractical for extensive population studies. For this reason, the method is decreasingly used. Nevertheless, a thorough assessment of variation over the entire mtDNA molecule can be obtained because the entire purified mtDNA can be digested directly with restriction enzymes. The digestion products (fragments) can be end-labelled with ^{32}P, examined by electrophoresis on either agarose or polyacrylamide gels, the gels exposed to X-ray film, and the results visualized directly from the autoradiographs. Additionally, highly purified mtDNA is required to manufacture mtDNA probes for Southern blot hybridization studies.

Total DNA extraction. Total DNA can be obtained with less time, expense and equipment, because the isolation procedure does not involve the separation of nuclear and mitochondrial DNAs. Tissues from almost any source (from white blood cells or buffy-coats to skin) can be used to isolate total DNA. In this procedure the whole tissue is ground to a powder in liquid nitrogen or with a tissuemizer (blood cells are simply isolated with low speed centrifugation), the separated cells are lysed by the addition of a detergent (sodium dodecyl sulphate) to release the DNA, the solution is extracted with phenol and chloroform, and the resultant DNA is precipitated with absolute ethanol and sodium acetate. Most of the step by step methods for isolating total DNA are modifications of the procedure described in Maniatis *et al.* (1989).

Methods of molecular analysis

There are two principal ways to uncover mtDNA sequence differences. The first involves direct nucleotide sequencing of regions in the mitochondrial genome; the second is indirect and involves the determination of restriction enzyme site variation. By far the most common approach has been the second, the comparison of restriction endonuclease digestion patterns within and between species, and this has been applied in a large number of studies on primates (Brown, 1981; Denaro *et al.*, 1981; Ferris *et al.*, 1981a,b; George, 1982; Johnson *et al.*, 1983; Wallace *et al.*, 1985; Hayasaka *et al.*, 1986, 1988b; Stoneking *et al.*, 1986, 1990; Cann *et al.*, 1987; Harihara *et al.*, 1988a,b; Schurr *et al.*, 1990). By contrast, there have been fewer nucleotide sequencing studies of mtDNA variation either within or between species of primates (Table 6.1).

Restriction enzyme analysis.   Restriction endonucleases are enzymes, naturally produced by a variety of prokaryotes, that can cut double-stranded DNA at specific sequences known as **restriction sites**. Restriction sites are generally four to six nucleotides long and have the characteristic of being palindromic (read the same in either direction). In a circular molecule like mtDNA, the number of fragments produced by a particular restriction enzyme is equal to the number of times the particular restriction site for that enzyme is present in the genome. These fragments can be separated by electrophoresis through a gel matrix. The image produced from a restriction enzyme analysis is a set of bands across an electrophoretic gel, with the largest fragments nearest the origin of sample application and the smaller fragments further away. The approximate size of these fragments can be determined using a known size standard. Bands migrating the same distance in different individuals are assumed to be bracketed by the same restriction sites, whereas fragment differences represent the unequal presence or absence of restriction sites. Restriction fragment profiles reflect the varied location of restriction sites throughout an individual's mtDNA (Fig. 6.2).

While considerable information can be obtained from restriction fragment data alone, a precise verification of the actual number of site similarities and differences requires a restriction site map. Restriction site mapping is a tedious process involving a series of digestions using different combinations of enzyme pairs (known as the **double-digestion** method: see Dowling *et al.*, 1990), but is more precise, because it reduces the probability of the incorrect identification of shared sites. For instance, fragments of similar length in two individuals may not be from the same region of the mitochondrial genome; therefore, comigration does not always mean similarity. In addition, in using fragment data, there is an increased likelihood that sequence divergence will be overestimated, because fragments differing in migration can still share some sites.

If purified mtDNA is used in the restriction enzyme reactions, the digested fragments are labelled with ^{32}P and examined by electrophoresis on either agarose or acrylamide gels. Radioactive labelling can be accomplished in several ways. In the case of purified mtDNA, restriction fragments are directly labelled by attaching radioactive nucleotides at one end, using a polymerase enzyme, before being electrophoretically separated. After separation, the gel is vacuum-dried and exposed to X-ray film. Very small fragments (less than 300 bp) can be visualized with this technique, thereby allowing the construction of a relatively high resolution map. In the detailed analysis of human mtDNA variation conducted by Cann *et al.* (1987) employing this technique, the resolution was even higher, because four-base site variation was compared directly to the nucleotide sequence of the entire human mitochondrial genome, allowing for exact mapping of most site variation.

If samples consist of total DNA, labelling requires the transfer of all the DNA from the electrophoretic gel to a membrane, usually made of charged nylon or nitrocellulose, and hybridization of separated restriction fragments to homologous pieces of radioactive mtDNA, called **probes**. This method is less sensitive because fragments smaller than 300–500 bp cannot be visualized with any degree of accuracy. As with the dried gel, the mtDNA fragments on the membrane are then visualized by autoradiography. In both cases, wherever a radioactively labelled fragment exists it will expose the film. These procedures are described in detail in Maniatis *et al.* (1989).

DNA probes can also be chemically labelled, thereby avoiding the potentially hazardous use of radioactive nuclides. Biotin and digoxygenin

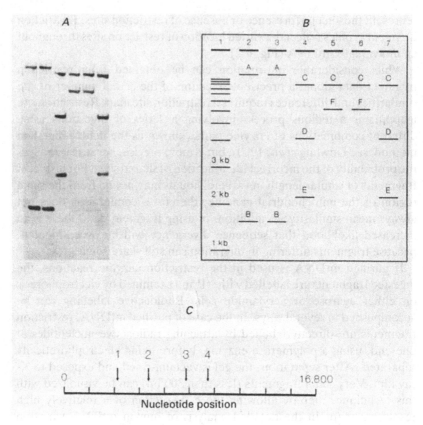

Fig. 6.2. Results of a restriction digest of macaque mtDNA with the enzyme Xba I. Lane 1 indicates the positions of the molecular weight markers from the 1 kb ladder (Gibco/BRL), which produces bands approximately every 1000 bp up to 12 000 bp. The rest of the lanes display haplotypes for *Macaca arctoides* (lanes 2 and 3), *M. assamensis* (lanes 4, 5 and 6) and *M. thibetana* (lane 7). *A*, Photograph of the original profiles following transfer to a nylon membrane, hybridization with a digoxygenin-labelled mtDNA probe and visualization with the non-radioactive method provided by the Genius kit (Boehringer Mannheim). *B*, Schematic view of the blot demonstrates more clearly the relationships of the haplotypes. Because the mtDNA molecule of each species is the same length, the fragments in each haplotype add up to the same total length. *C*, Restriction site map including all of the sites that produced the profiles in *A* and *B*. Fragment A is bracketed by site 4 on the left and site 3 on the right (remember that the molecule is actually circular). Fragment B, therefore lies inside sites 3 and 4. Fragment C, like fragment A, is bracketed by site 4 on the left, but ends at site 1 on the right. Consequently, fragment F lies inside sites 1 and 4. Fragments D and E result from the addition of site 2 in fragment F.

are two commonly used labels that allow target fragments to be visualized. The chemical reactions that produce the images involve immunological detection of the probe followed by either a colour reaction directly on the membrane or a chemiluminescent reaction in which the fragment profile is recorded on a piece of X-ray film.

Nucleotide sequencing. Direct nucleotide sequencing of mtDNA provides the most detailed data on differences among mtDNA haplotypes. The most commonly used method of nucleotide sequencing is the dideoxy chain termination procedure developed by Sanger *et al.* (1977). Sanger sequencing depends on the controlled enzymatic replication of a DNA fragment in which replication is terminated for a fraction of the molecules at each nucleotide position. Replication begins by annealing a primer (i.e. a short DNA sequence, typically 12–18 nucleotides long) to the fragment template. A polymerase enzyme then starts attaching free nucleotide bases, some of which are radioactively labelled, to the end of the primer. Elongation produces a sequence complementary to the template DNA. Some of the free nucleotides in solution are actually dideoxynucleotide analogs of the deoxynucleotides that constitute natural DNA. These analogs are incorporated into the growing string of nucleotides in the normal way, but once a dideoxynucleotide has been attached, that string will not be extended any further. Four separate reaction mixtures are made up, each containing one of the four dideoxynucleotides (ddATP, ddCTP, ddGTP, ddTTP). The newly synthesized DNA fragments from each reaction mixture are then separated by denaturing polyacrylamide gel electrophoresis and visualized by autoradiography. This form of electrophoresis allows one to distinguish between fragments that differ by a single nucleotide in length so that the profile allows one to reconstruct the sequence of bases directly from the autoradiograph (Fig. 6.3).

Sequence data contain far more detailed information than do restriction site data. Indeed, restriction site analysis can be viewed as the sequencing of very short pieces of DNA, the enzyme recognition sequences at each site, dispersed throughout the mitochondrial genome. In addition, analysis of restriction site data uses the presence or absence of sites, each containing about six bases, as phylogenetic characters, whereas every base is considered as a separate character when employing sequence data. With sequencing gels also the potential danger of misidentifying two non-homologous fragments as the same is avoided. However, restriction site analysis includes one important advantage for some applications; it embraces the entire genome, while direct sequencing

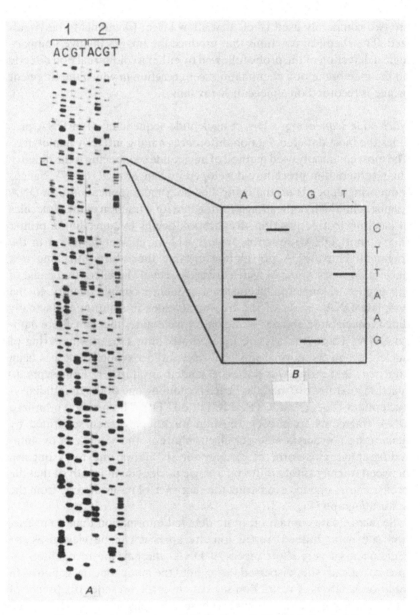

Fig. 6.3. *A*, A sequencing gel. *B*, A schematic view of the bracketed region illustrates more clearly the banding pattern and the way such an image is read.

examines only a limited region. The sequenced region may or may not reflect variation in the rest of the genome.

Polymerase chain reaction. The polymerase chain reaction (PCR) is a method that allows repeated enzymatic replication of short DNA sequences, generally under 4000 bases long (Innis *et al.*, 1990). This process is called amplification because from an initial very small quantity of DNA a large amount of the desired fragment is generated for sequencing or other types of genetic analysis. Short primer sequences flanking the desired fragment must be known *a priori* for most applications. PCR was made possible by the discovery of a bacterium that lives in association with oceanic thermal vents, *Thermus aquaticus*, and the subsequent purification of its DNA polymerase enzyme, Taq polymerase (Saiki *et al.*, 1988). This marked a breakthrough, because whereas previously known polymerases functioned only over a narrow temperature range, Taq polymerase is stable over a very wide range of temperatures, including temperatures approaching 100 °C in which DNA is denatured.

A typical PCR reaction involves a three-stage cycle. (1) A high temperature stage (about 95 °C) denatures the DNA and separates the primers from the template. (2) A relatively cool stage (about 50 °C) allows the primers to anneal to the template. (3) A stage of intermediate temperature (about 72 °C) allows the Taq polymerase to build the sequence between the primers. Because a new set of templates is generated each cycle, the amount of target DNA increases almost exponentially. The cycle is usually repeated 35–45 times, thus amplifying the template as much as 10^5-fold in a matter of hours (Saiki *et al.*, 1985, 1986).

Because this method is so rapid (Allard *et al.*, 1991), it has replaced the use of cloning (see Maniatis *et al.*, 1989) to obtain highly purified quantities of specific portions of the mtDNA genome. In combination with nucleotide sequencing and restriction enzyme analysis, PCR has allowed researchers to generate a great deal of mtDNA character data for a variety of primate species, which have then been applied to phylogenetic questions (e.g. Vigilant *et al.*, 1989; Ruvolo *et al.*, 1991). Because so little template DNA is required for PCR amplification, DNA from previously unusable sources, such as single hairs, preserved museum specimens, archaeological remains and even fossil material, can now be studied (Higuchi *et al.*, 1988; Pääbo *et al.*, 1988; Pääbo, 1989; Vigilant *et al.*, 1989; Thomas *et al.*, 1990). The primary drawback of the technique, as with cloning, is that the information gained is limited to one small portion of the mitochondrial genome.

Quantifying mtDNA variation

The methods described above provide three kinds of data: restriction fragment data, restriction site data, and nucleotide sequence data. All three types of data can be used to determine how different two mtDNA molecules, derived from two individuals, are from one another. More specifically, one can determine the number of nucleotide substitutions that have occurred since these two individuals last shared a common matrilineal ancestor (i.e. the same mtDNA molecule).

Estimating divergence from restriction enzyme analyses. Several methods have been developed for estimating overall sequence divergence (genetic distance) between mtDNA haplotypes from both restriction enzyme site and fragment data (Upholt, 1977; Nei and Li, 1979; Gotoh *et al.*, 1979; Engels, 1981; Nei and Tajima, 1981). The logic behind these methods is similar in that the amount of nucleotide sequence divergence between two mtDNA haplotypes is correlated with the proportion of shared fragments or sites. For this reason all approaches yield similar estimates (Kaplan, 1983), and thus we recommend the method of Nei and Li (1979), because its mathematical logic is straightforward, its calculation is simple (see Nei, 1987: 106–7) and most of the divergence estimates in the current literature have been derived using this technique.

In the method of Nei and Li (1979), nucleotide sequence divergence, d, between two haplotypes can be calculated from restriction site data using their Equations 9 and 10, and from fragment data using their Equations 20 and 21 on the assumption that fragments of similar mobility share homologous sites. A more precise, maximum likelihood estimate of d can be made when restriction site data are available, using the algorithm of Nei and Tajima (1981, 1983). Estimates of d should be calculated separately for four-base and six-base enzymes and the values weighted relative to the number of cleavage sites produced by each set (Nei, 1987).

Aside from the standard estimates of haplotype divergence calculated using the method of Nei and Li (1979), one also can estimate haplotype or nucleotide diversity (Nei and Tajima, 1983), a measure equivalent to nuclear DNA diversity (Nei, 1973). A simple description of this method, including an application to human mtDNA data, can be found in Nei (1982). These population statistics can prove valuable when conducting population-level and geographic studies, as well as when trying to estimate net sequence divergence of two taxa since they became reproductively isolated (Nei, 1987).

Estimating divergence from direct nucleotide sequencing. The most direct measurement of divergence can be obtained from nucleotide sequences. Although the most straightforward approach for estimating divergence is a direct count of nucleotide substitutions separating a pair of individuals, most researchers advocate the incorporation of correction factors in a quantitative estimate of sequence divergence. These estimates provide corrections for silent and replacement site substitutions, transition/transversion bias, and the increased probability of multiple substitutions with increasing evolutionary divergence (Jukes and Cantor, 1969; Kimura, 1980; Miyata and Yasunaga, 1980; Brown *et al.*, 1982; Gojobori *et al.*, 1982a,b; Tajima and Nei, 1984). Each correction, however, assumes a particular model for molecular evolution, and these assumptions should be evaluated prior to applying the model. In some cases, the correction factor probably is not valid for mtDNA sequence divergence (e.g. Jukes and Cantor, 1969).

When comparing non-protein coding genes such as the rRNA genes and the D-loop region, sequence alignment can become difficult because additions and deletions of bases rather than base substitutions may occur at a high frequency. Although no multiple sequence program is guaranteed to provide the correct alignment, there are several types that can be used (e.g. DNA-Star).

Estimating evolutionary rates. Prior to estimating divergence times from molecular data, rates of evolution should be examined. Several methods can be used for this purpose.

(1) *Relative Rate Test* (Sarich and Wilson, 1967). This test for rate uniformity is one of the most broadly used in molecular evolution. It requires no knowledge of divergence times between species, but does presuppose branching order in that an outside reference species or outgroup is required for the examination of lineages sharing a common point of divergence. The test is a comparison of magnitude of change along the two lineages subsequent to divergence from a common ancestor and relative to the outside reference. More than one outside species should be used in order to minimize the effects of back mutations and parallel and convergent substitutions (Beverley and Wilson, 1984). The effects of such homoplasy obviously increase over evolutionary time, so that several calibration points are needed (Gingerich, 1986). This test does not assess rate constancy, only colinearity of rates in different branches of a tree (Gingerich, 1986).

(2) *Star Phylogeny Approach* (Kimura, 1983). This test considers a case where all species diverged at the same time from a common ancestor,

and compares the observed and expected variances in rate under a Poisson process. This approach may be valid for mammalian orders, but the estimates are probably minimal as a result of dichotomous branching (Nei, 1987). Gillespie (1986) has modified this approach.

(3) *Langley and Fitch* (1974) Method. This procedure requires knowing the branching order whereupon expected branch lengths are calculated using maximum likelihood and rate heterogeneity is tested using a χ^2.

(4) *Absolute Rate.* This approach requires a phylogenetic tree, an estimate of substitutions along each branch, and an evolutionary time estimate or calibration point derived from either fossils or biogeography. One can then relate divergence time between two species to the extent of divergence separating the species.

Phylogenetic reconstruction. The phylogenetic relationships among mtDNA haplotypes or sequenced fragments are derived using one of two approaches, phenetics or cladistics (Felsenstein, 1982, 1988; Miyamoto and Cracraft, 1991). The phenetic approach requires the calculation of a distance estimate from restriction fragment, restriction site, or nucleotide sequence data, followed by a phylogenetic reconstruction based on levels of divergence observed between pairs of taxa. Estimates such as *d* are used quite often in phenetic comparisons of mtDNA haplotypes, and corrected nucleotide sequence data can similarly be used. Some of the more common procedures for tree construction based on phenetic distances are UPGMA (Sneath and Sokal, 1973), Fitch–Margoliash (Fitch and Margoliash, 1967), and Neighbour-Joining (Saitou and Nei, 1987). UPGMA is the most problematic algorithm, because it assumes a constant rate of nucleotide substitution, and does not consider convergent or parallel evolution, so that phylogenetic hypotheses are difficult to evaluate. The other two methods do not have these shortcomings, but the calculation of any corrected distance requires some assumptions about how evolution proceeds.

The cladistic approach to phylogeny reconstruction (Hennig, 1966) is based on character state data or the actual uncorrected fragments, sites, or nucleotide sequences, and phylogenetic trees are constructed using some form of maximum parsimony. The only requirement for choosing among phylogenetic trees derived from parsimony analysis is that the tree with the fewest number of evolutionary events (distribution of character changes throughout the tree) is the one chosen to explain the data (Swofford and Olsen, 1990). A parsimony analysis begins by scoring sites, fragments, or sequences as either present or absent followed by a phylogeny using one of several computer programs: PAUP 3.0 (Swofford, 1990); HENNIG86 (Farris, 1988), MacClade (Maddison and Maddison, 1992); PHYLIP (Felsenstein, 1990). In the case of nucleotide sequence data, there are approaches for conducting weighted parsimony (Williams and Fitch, 1989; Wheeler, 1990; Swofford, 1990). All of these approaches are attempts to correct for biases in the sequence data associated with disproportional transitions versus transversions, differential change at first, second and third positions, and skewed changes involving particular bases.

Finally, several recent studies using parsimony analysis have advocated estimates of confidence limits for particular branching patterns. These methods are generally a form of randomization or bootstrapping and can be found in Felsenstein (1989). These approaches, however, are limited by the number of replications that can be formed and the number of taxa included in an analysis. Thus, these estimates do not reflect statistical confidence in a true inferential sense, but yield an empirical estimate of the probability of getting a particular tree given some sample of the possible trees that could be formed with the existing character state data.

Application to evolutionary research

Studies of mtDNA variation in primates have focused on both intraspecific and interspecific levels of divergence. The most detailed studies of intraspecific variation have been conducted on cercopithecoid primates and human populations, while most comparative studies of interspecific divergence pertain to the phylogenetic relationships among hominoid primates, particularly *Gorilla*, *Pan* and *Homo*.

Population genetics

With rare exceptions, most population genetic research on non-human primates has been conducted on Old World monkeys, particularly the cercopithecines (Turner, 1981; Shotake, 1981; Nozawa *et al.*, 1982; Melnick *et al.*, 1984a,b). Additionally, very few studies of primate mtDNA have either determined or used population genetic data to explore the microevolutionary dynamics of the mitochondrial genome (Table 6.2). For this reason much of the following section will focus on the research of the authors, and include limited data from other studies.

Cercopithecoids. In many cercopithecine monkey species, virtually all males leave the social group into which they are born, while females

Table 6.2. *Intraspecific mtDNA variation in primates*

Taxa	% Sequence divergence	Reference
Cercopithecoids		
Macaca assamensis	0.2–6.0	Hoelzer et al., unpublished data
M. fascicularis	0.3–2.1	Harihara et al., 1988a
	–4.1	George, 1982
M. fuscata	0.8–2.3	Hayasaka et al., 1988b
M. maurus	0.6–1.0	Williams, 1990
M. mulatta	0.2–4.5	Melnick et al., unpublished data
	0.5–2.0	Zhang and Shi, 1989
M. nemestrina	1.0–8.4	Williams et al., unpublished data
		Rosenblum et al., 1992
M. sinica	–3.1	Hoelzer et al., unpublished data
M. tonkeana	–3.6	Williams, 1990
Hominoids		
Pan troglodytes	0.5–2.0	Ferris et al., 1981a
P. paniscus	–1.5	Ferris et al., 1981a
Gorilla gorilla	0.3–0.9	Ferris et al., 1981a
Pongo pygmaeus	0.5–5.0	Ferris et al., 1981a
Homo sapiens	0.1–1.3	Cann et al., 1987

remain in their natal group throughout their life (Sade, 1972; Pusey and Packer, 1987; Clutton-Brock, 1989; see Moore, 1984 for exceptions). Given the extreme sex-biased asymmetry in dispersal among these primates, it is reasonable to expect the distribution of genetic variation, especially with respect to mtDNA, to reflect these demographic patterns.

In cercopithecine monkey species that exhibit a social system with extreme female sedentism and complete male dispersal, observed patterns of nuclear variation within a population are as follows (Melnick et al., 1984a; Melnick, 1988): (1) most of the genetic variation (92–97%) in a local population can be found in any one social group and (2) only a small amount of variation (3–8%), though statistically significant, can be attributed to differences between groups.

At the regional level, there is a similar pattern to that seen in populations, with estimates of 90–93% of a region's diversity apportioned to differences between individuals in the same local population. Major differences, however, do appear at the species level, with species having geographic ranges fragmented by water barriers exhibiting much more structure relative to the overall pattern of nuclear gene variation. For example, only 67% of the species' diversity is found in the regional populations of the crab-eating macaque (*Macaca fascicularis*). When major geographic barriers are absent more than 90% of a species' diversity can be attributed to differences between individuals within a single region (Melnick, 1988).

In summary, the general pattern of overall nuclear gene variation in macaque populations consists of a background of broad genetic homogeneity, caused largely by high rates of male migration and gene flow, against which one also finds small, statistically significant genetic differences between groups, attributable primarily to differences between each group's sedentary adult females and their offspring (D. J. Melnick *et al.*, unpublished data). Where information is available (see Melnick, 1987, 1988), a macaque-like pattern of population genetic structure is also found among the African cercopithecines (i.e. baboons and vervets). Given the ubiquity of male dispersal in higher primates (Pusey and Packer, 1987; Clutton-Brock, 1989), it is likely that nuclear population genetic structure of most Old World monkey species resembles what is seen in the macaques.

In contrast to the relatively homogeneous distribution of nuclear gene variation, one would expect maternally inherited genetic variation to be far more differentiated between groups and populations as a result of female sedentism among macaques. This is precisely what one finds (Table 6.2). Rhesus monkey populations exhibit exceptionally high levels (as great as 4.5%) of estimated mtDNA sequence divergence (D. J. Melnick *et al.*, unpublished data). Quantitative analysis (Nei, 1982) indicates that roughly 9% of the total species diversity can be apportioned to within population differences, whereas a full 91% consists of differences between populations (Fig. 6.4). These proportions are the exact opposite of what is found in the nuclear genome (Melnick and Hoelzer, 1992).

Data on mtDNA variation in other cercopithecine primate species, though limited, corroborate the findings on the rhesus monkey. Among populations of *M. fuscata*, sequence diversity was distributed solely as interpopulational differences (Hayasaka *et al.*, 1986). Similarly, average sequence divergence within populations of *M. fascicularis* was 25% of that found between populations (Harihara *et al.*, 1988a). This pattern was also found in *M. nemestrina* (A. K. Williams *et al.*, unpublished data), where differences between populations were on average an order of magnitude greater than those within populations.

In more detailed studies of local populations, either no mtDNA variation was found (e.g. *M. mulatta*), or differences were distributed solely between social groups (e.g. *M. sinica*) (D. J. Melnick *et al.*, unpublished data; G. A. Hoelzer *et al.*, unpublished data). Hence, members of any social group were monotonously uniform in their

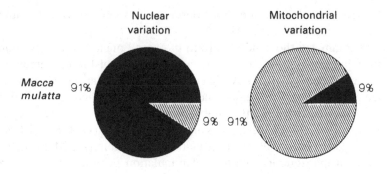

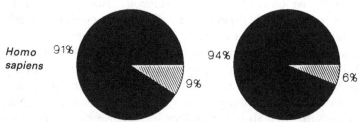

Fig. 6.4. Distribution of genetic variation in the rhesus monkey (*M. mulatta*) and humans (*H. sapiens*). Hatched areas indicate variation between geographic regions: solid black areas indicate variation contained within regions. Note that, unlike human nuclear and mtDNA and rhesus nuclear DNA, the mtDNA variation found in rhesus monkeys is highly structured among geographic regions.

mtDNA haplotypes, and all intrapopulational differences were distributed as between-group differences. Again, this is in marked contrast to the distribution of nuclear genetic variation in these same local populations (Melnick *et al.*, 1984a; D. J. Melnick, unpublished data).

It is clear from analyses of rhesus monkeys and other macaque species (e.g. *M. fascicularis*, *M. fuscata*, *M. nemestrina*, *M. sinica*) that the population structures of nuclear and mitochondrial genetic variation are very different. With the exception of major water barriers, nuclear genetic variation in macaques is quite uniformly distributed throughout a species' range. Thus, for example, most of the rhesus species diversity can be found in any single regional subdivision and the differences between rhesus populations are surprisingly low, given the geographic distances that separate populations in this widely distributed species.

The distribution of mitochondrial genetic variation is the mirror image of nuclear genetic structure. Almost all of a species' diversity is distributed as between population differences and there is relatively little within-group or within-population diversity. Hence, depending upon which genome is used to characterize the rhesus, completely different pictures of population genetic structure are obtained.

For some time it has been clear that the mitochondrial genome is a rich, easily interpreted source of genetic variation for population genetic analysis (Avise, 1986; Avise *et al.*, 1987). If the population structures of the two genomes are similar, mtDNA will in most cases provide a simpler and more abundant source of data with which to analyse the population genetics of Old World monkeys. However, current research indicates that the population structure of mitochondrial genetic variation is not merely a poor reflection of nuclear population genetic structure, but is in fact a positively misleading one. Large mtDNA differences between populations may be associated with large or with small nuclear genetic differences. In macaques, the latter case is most common. While the population genetics of mtDNA are intrinsically interesting (Avise, 1986), the mitochondrial genome represents such a small fraction of an organism's overall genetic make-up and its population genetic structure in macaques is so different from the nuclear genome, that it tells us virtually nothing about the overall genetic similarities or differences within and between macaque populations.

Hominoids.   Essentially, no data on nuclear gene variation exist for the hylobatids or pongids. Thus, whether the apes share the macaque-like population structure or not is largely unknown. In contrast, a great deal of attention has centred on the genetic structure of the human species. The results of these studies indicate that nuclear population genetic structure roughly parallels the pattern described for the monkeys, with most (*c.* 90%) of the species' diversity attributable to differences between individuals in the same geographical 'race' and very little of the overall diversity (*c.* 10%) attributable to interracial differences (Lewontin, 1972; Nei, 1982; but see Smouse *et al.*, 1982). Again, it is assumed that these broad similarities reflect persistent, substantial gene flow between the major geographical regions of the range of *Homo sapiens* (Lewontin, 1972), though they may also reflect descent from a relatively recent common ancestor (Nei, 1982).

The analysis of mtDNA in the hominoids has, with the exception of humans, generally been restricted to single individuals from each species with the aim of clarifying phylogenetic relationships (e.g. Brown *et al.*,

1979; Ferris *et al.*, 1981a; Hixson and Brown, 1986). Nevertheless, some intraspecific sampling has been done (Ferris *et al.*, 1981b). These studies have revealed a considerable amount of variation (sequence divergence) between conspecifics, ranging from about 0.5% among lowland gorillas to 5.0% between Bornean and Sumatran orangutan haplotypes (Table 6.2). The population genetic structures of these species are not presented in the studies cited here. However, published data indicate that even though haplotypes tend to be localized, there can be considerable variation among haplotypes in the same region (e.g. Sumatran orangutans). Unfortunately, because the exact provenance of the samples used in the study of Ferris *et al.* (1981b) is not known, a detailed population genetic analysis of these data is not possible. Additionally, the lack of a parallel set of data for nuclear gene variation in hominoids (excluding humans) prohibits a comparison of the patterns of nuclear and mitochondrial DNA variation.

Human mtDNA has been studied in great detail, and many mtDNA haplotypes have been found throughout Old World and New World populations (Brown, 1980; Denaro *et al.*, 1981; Johnson *et al.*, 1983; Horai *et al.*, 1984; Wallace *et al.*, 1985; Brega *et al.*, 1986a,b; Horai and Matsunaga, 1986; Stoneking *et al.*, 1986, 1990; Cann *et al.*, 1987; Harihara *et al.*, 1988b; Santachiara-Benerecetti *et al.*, 1988; Vigilant *et al.*, 1989; Pennington *et al.*, 1989; Schurr *et al.*, 1990; Vigilant *et al.*, 1991). Although earlier studies of mtDNA restriction site variation revealed differences between some geographic populations of humans (Brown, 1980; Johnson *et al.*, 1983), many mtDNA haplotypes were found to be shared among divergent human groups (Cann *et al.*, 1987). Surprisingly, these haplotypes differ very little from one another, with sequence divergence ranging from 0.00 to 1.3% and averaging 0.32% (Cann *et al.*, 1987). Even more surprising, the level of intrapopulation variation is roughly equal to interpopulation variation. Hence, most of the species diversity in mtDNA (*c.* 94%) can be found within any one major geographical population (i.e. race) and very little (*c.* 6%) consists of differences between populations (Cann *et al.*, 1987; Whittam *et al.*, 1986; Nei, 1982, 1985). While this parallels very closely what one finds in the nuclear genome of this species (Fig. 6.4), it stands in sharp contrast to the known mtDNA population structures of macaques and the apparent population structures of the other hominoids. On the face of it, human mtDNA population structure seems to indicate the historical presence of significant female dispersal, either individually or with larger kin groups.

Recently, more detailed studies of geographic variation in specific aboriginal human populations, using both high resolution restriction site mapping (Stoneking *et al.*, 1990) and nucleotide sequencing of the D-loop region, the fastest evolving component of the mitochondrial genome (Vigilant *et al.*, 1989), have painted a different picture. The nucleotide sequence information, in particular, reveals considerable geographic variation, with !Kung populations exhibiting low within-population variation and high between-population differences relative to Western and Eastern Pygmies and other human populations. In the !Kung, there is a correlation between shared mtDNA haplotypes and geographic proximity, suggesting that female lineages of hunter–gatherers move only short distances. Among Pygmies, the Eastern and Western populations share most alleles at the 44 nuclear gene loci examined, while they share no mtDNA haplotypes. These data also suggest that female hunter–gatherers may be much more sedentary than their male counterparts. To this point, patterns of mtDNA variation indicate that among the remaining small aboriginal human populations genetic structure and levels of female dispersal are very similar to what is seen in macaque populations. In contrast, the more widely distributed human groups (e.g. Bantu, European, East Asian) reveal an overall pattern of considerable mtDNA admixture and significant female dispersal, in all likelihood reflecting their recent colonization of vast geographic areas. These contrasting patterns seem to suggest that the aboriginal population genetic data are more likely to reflect both the mtDNA population structure and the levels of female dispersal that persisted throughout most of human evolution. Hence, the macaque model may be broadly applicable to the evolution of all Old World higher primates, including hominids.

Conclusion. As we have stated, the population genetic structure of macaque mtDNA variation is the converse of the distribution of nuclear genetic variation. This lack of congruence may be attributable to both the population biology of male and female macaques and the unique transmission genetics of mtDNA. However, regardless of its cause, mtDNA remains a questionable population genetic tool in these species, because it is likely to exaggerate differences between populations that are genomically quite similar. The nuclear genome, on the other hand, incorporates most of an organism's genetic material and thus, if neutral or nearly neutral nuclear gene loci are used (Nei, 1987), an accurate picture of overall population genetic structure and interpopulation differences can be obtained. Computer simulation studies (G. A. Hoelzer *et al.*, unpublished data) and what is known of primate dispersal patterns (Pusey and Packer, 1987) suggest that a macaque-like mtDNA population structure is likely to be found in many or most Old World monkeys.

If this turns out to be true, similar limitations on the use of mtDNA in the population genetic analyses of these species will apply.

Despite these drawbacks, there are some population genetic investigations that can be enhanced by using mtDNA. These include (1) tracing the geographic origin of migrant males, (2) reconstructing a matrilineal phylogeny of social groups within a population and (3) determining the direction of gene flow between major geographical populations or closely related species in zones of contact. The first two applications can be combined to examine whether rates of male exchange are positively or negatively correlated with maternal relatedness between groups. The latter application will allow one to determine whether particular individual or sex-specific behaviour patterns influence the direction of gene flow (J. Supriatna *et al.*, unpublished data). All of these uses of mtDNA data are possible because females generally do not migrate and because the maternal inheritance, lack of recombination and rapid evolution of mtDNA lead to local homogeneity and large interpopulation differences. Hence, mtDNA haplotypes are excellent indicators of maternal relatedness and geographic origin.

Across major geographic regions, human mtDNA population structure parallels human nuclear population genetic structure and therefore can be used as a substitute. Because the mitochondrial genome evolves at a rate five to ten times as fast as the nuclear genome (Brown *et al.*, 1979), one generally finds much more variation, and thus genetic information, when comparing closely related populations or species. Given the greater expected amount of genetic information, mtDNA may in fact be a superior genetic tool for investigating the regional population structure of human genetic variation. Within a region, however, aboriginal mtDNA population structure follows a macaque pattern, exhibiting large interpopulation differences. These differences do not parallel the nuclear population genetic structure, one that reflects broad similarities between these populations. Thus, as in the case of macaques, the distribution of human mtDNA does not accurately reflect the overall pattern of genomic similarities and differences in aboriginal human populations. The utility of mtDNA in the study of the population genetics of other hominoids cannot be assessed without further study.

Phylogenetic reconstruction

Given the simple transmission genetics of mtDNA (see above) and the statistical process of lineage sorting (Avise *et al.*, 1984; G. A. Hoelzer *et al.*, unpublished data) it is logical to assume that the differences seen between populations are primarily caused by the fixation of mutations in specific matrilines, rather than recombination of genetic material. In macaques and most higher primates, nearly complete female philopatry removes the effects of the dispersal of mtDNA haplotypes from one population to another, and thus simplifies the interpretation of differences even further. This, of course, would not be the case for nuclear variation, where recombination and migration are important determinants of interpopulation differences. Additionally, the rapid nucleotide substitution rate in mtDNA (Brown *et al.*, 1979) and the differential rates of divergence seen for various components of the mitochondrial genome provide a large number of phylogenetically informative characters that can be used in analyses of intraspecific and interspecific patterns of variation. For this reason, while mtDNA data have certain limitations in population genetic analysis, they may provide a superior source of information for phylogenetic reconstructions.

Cercopithecoids. Most phylogenetic studies of Old World monkeys have used either restriction enzyme analysis (George, 1982; Hayasaka *et al.*, 1988b; Williams, 1990; D. J. Melnick *et al.*, unpublished data; G. A. Hoelzer *et al.*, unpublished data) or nucleotide sequences (Hayasaka *et al.*, 1988a; Disotell *et al.*, 1992). The most thorough restriction enzyme studies have focused on macaques, with few exceptions (George, 1982). The nucleotide sequencing studies have examined variation in three regions of the mitochondrial genome: (1) an 896 bp region (also sequenced for hominoid primates by Brown *et al.*, 1982) containing three tRNA genes (tRNA^{His}, tRNA^{Ser}, tRNA^{Leu}) plus portions of both the 3' region of the NADH-dehydrogenase subunit 4 (ND-4) and the 5' region of ND-5 (Hayasaka *et al.*, 1988a); (2) the cytochrome *c* oxidase subunit II (COII) gene, a 684 bp region encoding a small subunit of the cytochrome oxidase complex which interacts with nuclear cytochrome *c* during electron transport (Ruvolo *et al.*, 1991; Adkins and Honeycutt, 1991; Disotell *et al.*, 1992); and (3) a 430 bp segment of the 12S ribosomal RNA gene (D. J. Melnick *et al.*, unpublished data).

Higher level systematics

Hayasaka *et al.* (1988a) derived a mtDNA phylogeny for a number of divergent primate taxa, including some Old World monkeys (Fig. 6.5A), using the nucleotide sequences of the 896 bp region, the neighbour-joining method (Saitou and Nei, 1987), and midpoint rooting. The overall branching pattern of the Hayasaka tree is consistent with conventional

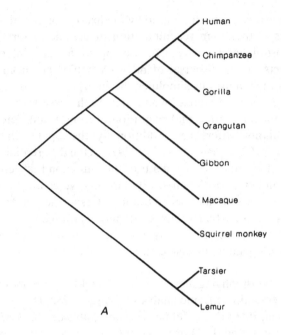

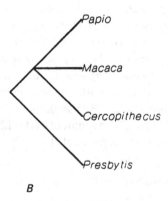

Fig. 6.5. *A*, Phylogenetic tree of primate groups based on sequencing of an 896 bp fragment of mtDNA (Brown *et al.*, 1982; Hayasaka *et al.*, 1988a). *B*, Consensus tree of cercopithecoid clades based on restriction site maps (George, 1982).

ideas on primate relationships (Napier and Napier, 1967) as well as with recent palaeontological (Fleagle, 1988) and molecular (Goodman, 1976) data. A closer look at the tree also reveals a macaque phylogeny, which places the Japanese, rhesus and crab-eating macaques in a group distinct from the Barbary macaques, a configuration consistent with palaeontological (Delson, 1980), morphological (Fooden, 1980), and blood protein (Melnick and Kidd, 1985; Fooden and Lanyon, 1989) analyses.

Mitochondrial DNA restriction site data are more equivocal with respect to Old World monkey relationships (Fig. 6.5*B*; George, 1982). The resultant phylogeny is not congruent with morphological and palaeontological data (Szalay and Delson, 1979). While colobines and cercopithecines are distinguished, it is not clear from the restriction site data whether the papionins (*Papio* and *Macaca*) form a monophyletic clade distinct from *Cercopithecus*. A consensus tree constructed from the five best Wagner trees (Farris, 1970) for George's (1982) data consists of a trifurcating cluster of *Papio*, *Cercopithecus* and *Macaca*. The tree based on the conventional taxonomy of the cercopithecines, however, requires only three additional mutational events when compared with the most parsimonious tree that groups *Cercopithecus* and *Macaca* together with *Papio* being further out.

A recent study by Disotell *et al.* (1992) examined nucleotide sequence variation in the COII gene of *Macaca* (macaques), *Papio* (baboon), *Mandrillus* (drill), *Theropithecus* (gelada), *Cercocebus* (mangabeys), and *Cercopithecus* (vervet). The important conclusions to be drawn from this study are: (1) *Theropithecus* and *Papio* form a clade, (2) *Mandrillus* is more closely related to *Cercocebus* than it is to *Papio* and (3) the most parsimonious tree suggests that the macaques represent a separate radiation from the exclusively African Papionini, with mangabeys plus drills and baboons plus gelada splitting at a later date. The relationships, however, among the macaque, baboon, and *Mandrillus/Cercocebus* clade relative to each other and to *Cercopithecus* are equivocal. Overall, the phylogenetic tree derived from this mitochondrial gene is congruent with other molecular data but in opposition to phylogenies derived from morphological characters. These data in combination with George's (1982) restriction site data were unable to resolve convincingly the relationships among *Cercopithecus*, *Papio* and *Macaca*.

Relationships among species of Macaca

Looking even more closely at the cercopithecines, there exists a considerable amount of mtDNA data on the macaques (Hoelzer and Melnick,

1992). The genus *Macaca* consists of 16–19 species (Fooden, 1980; Groves, 1980) occupying a vast range from eastern Afghanistan in the west to Japan, Taiwan, the Philippines and Sulawesi in the east, and is generally subdivided further into three or four species groups (Fooden, 1980). The mtDNA for members of each one of these species groups has now been examined, using restriction fragment, restriction site, or nucleotide sequence data. In some cases more than one type of data has been used to reconstruct the phylogeny of the group.

The *fascicularis* group, which consists of the rhesus, Japanese, Formosan, and crab-eating (long-tailed) macaques has been the most extensively studied. Recently, the restriction sites of 15 enzymes for 12 unique rhesus haplotypes have been mapped (Melnick *et al.*, 1992). Comparing these maps with those reconstructed from data presented by Hayasaka *et al.* (1988b), a phylogenetic tree for all the members of the species group was constructed (Fig. 6.6A). This analysis demonstrated two important points about the use of intraspecific variation in interspecific comparison. First, when considerable intraspecific variation exists, as it does for many of the macaque species examined thus far, failure to consider this variation in calculations of interspecific differences leads to an overestimation of differences and divergence times separating taxa. Secondly, mtDNA is an excellent phylogenetic tool for macaques, because it preserves a record of biogeographic and cladogenic events that have been obscured in the nuclear genome by high rates of male-mediated gene flow and recombination. In the case of the *fascicularis* group, we discovered a major east–west mtDNA discontinuity in the rhesus monkey and a paraphyletic origin of Japanese macaques, and possibly Formosan macaques, from the eastern rhesus branch (Fig. 6.6A). Hence, the Japanese and eastern rhesus macaques share a closer common mitochondrial ancestor than do the rhesus monkeys in the eastern and western parts of their range. Evidence of the split within the rhesus species was probably obscured in the nuclear genome (Melnick and Kidd, 1985; Fooden and Lanyon, 1989) by extensive gene flow through male migration subsequent to the initial divergence.

A third, and perhaps equally significant, result of the macaque data is that intraspecific variation can also have an effect on the reconstruction of interspecific phylogenetic relationships. This result advises caution when constructing mtDNA phylogenies among closely related taxa using single representatives of a species because gene trees may not always reflect species trees when intraspecific and interspecific levels of divergence are similar (Takahata, 1989). The general topology of our *fascicularis* tree is otherwise consistent with a similar restriction site analysis on a more

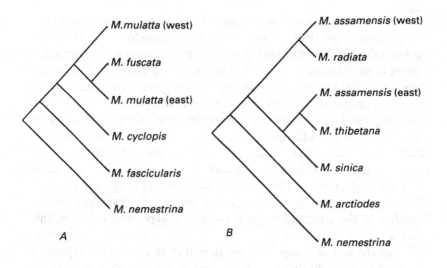

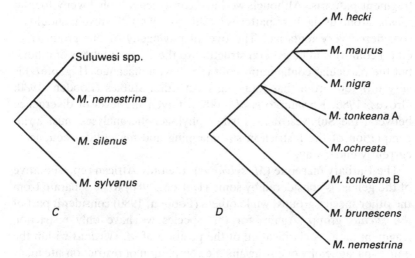

Fig. 6.6. Phylogenetic trees of species groups in the genus *Macaca*. The *fascicularis* group tree (*A*) is based on restriction site maps, whereas the trees for the *sinica* group (*B*), the *silenus–sylvanus* group (*C*) and the Sulawesi macaques (*D*) are based on restriction fragment profiles.

limited sample of individuals (Hayasaka *et al.*, 1988b) and nucleotide sequence data on three of the four constituent species (Hayasaka *et al.*, 1988a).

In many ways, determining the relationships among species in the *sinica* group of macaques has presented the most problems for systematists (Fooden, 1980, 1990; Delson, 1980). One reason for these difficulties has been the highly derived set of characters possessed by *M. arctoides*, leading to the variable placement of *arctoides* as a member of the *sinica* clade, as a sister taxon to the clade, or as a completely separate, distinct species group (Fooden, 1990). Our analyses of this group (G. A. Hoelzer *et al.*, unpublished data) reveal extraordinary amounts of intraspecific variation in several of the group's constituent species (e.g. *M. sinica*, *M. assamensis*), and suggest additional evolutionary complexities in the macaque phylogeny (Fig. 6.6*B*), which had previously gone unnoticed in the analysis of allozymes (Melnick and Kidd, 1985; Fooden and Lanyon, 1989). In particular, *M. assamensis* haplotypes fall in two very distinct branches of the tree, indicating a possible polyphyletic origin of this species.

Within the *silenus* group, only the Sumatran *M. nemestrina* haplotype has been completely mapped. Partial maps for five other *nemestrina* haplotypes exist. The rest of the data for this group consists of restriction fragment patterns. Although we have constructed a phylogeny for the *silenus* group using these patterns (Williams, 1990), the results should be considered very tentative. The overall phylogeny of this group (Fig. 6.6*C*) conforms in its general structure to the reconstructions of others, but the particular configuration of the Sulawesi macaques (Fig. 6.6*D*) is very different from that found in most other studies (Fooden, 1980; Groves, 1980; Kawamoto *et al.*, 1982). Given the level of divergence between the Sulawesi taxa, further phylogenetic analyses must await completion of the restriction site mapping and nucleotide sequencing currently under way.

The Barbary macaque (*M. sylvanus*), the only African representative of the genus, is considered by some (Delson, 1980) to be separate from the other species groups, while others (Fooden, 1980) consider it part of the *silenus* group. Again, for this species we have only restriction fragment data. A clarification of the position of *M. sylvanus* within the entire macaque phylogeny awaits the completion of restriction site maps for *sylvanus*, the other unmapped macaque species, and an outgroup (i.e. *Papio cynocephalus*).

Hominoid primates. The derivation of phylogenetic relationships among the major hominoid genera, especially *Homo*, *Pan* and *Gorilla*, has been a major focal point for many systematists, and most recently two opposing views by molecular systematists have resulted in a rather heated debate (Sibley and Ahlquist, 1984, 1987; Caccone and Powell, 1989;

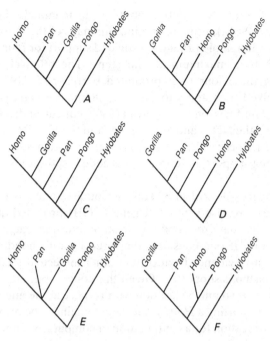

Fig. 6.7. Alternative phylogenetic relationships that have been variously derived for homonoids.

Sarich *et al.*, 1989; Sibley *et al.*, 1990). The debate has focused on which of several phylogenetic hypotheses (Fig. 6.7) is best supported by the existing comparative data, including morphology, chromosomes, amino acid sequences, nucleotide sequences, and DNA–DNA hybridization. Morphological characters can be found to support any one of several hypotheses (Andrews, 1987), with most studies favouring a *Pan/Gorilla* clade to the exclusion of *Homo*. Most molecular studies (with the exception of Djian and Green, 1989) reveal two patterns: (1) A large amount of change supporting *Homo/Pan* to the exclusion of *Gorilla*, suggested by DNA–DNA hybridization and nucleotide sequences from nuclear ribosomal RNA genes and the mitochondrial COII gene (Sibley and Ahlquist, 1984, 1987; Sibley *et al.*, 1990; Caccone and Powell, 1989; Ruvolo *et al.*, 1991); (2) little, if any, support for a *Homo/Pan* clade (Sarich and Wilson, 1967; Ferris *et al.*, 1981a; Brown *et al.*, 1982; Goodman *et al.*, 1983; Hixson and Brown, 1986; Koop *et al.*, 1986; Miyamoto *et al.*, 1987; Hayasaka *et al.*, 1988a).

Earlier approaches to resolve the *Homo, Pan, Gorilla* relationships with mtDNA analysis did not yield a definitive solution, despite the accelerated rate of mtDNA evolution observed in mammals. For instance, Ferris *et al.* (1981a), using restriction site maps of the mitochondrial genomes of six hominoid primates, provided support for four trees, differing in length by a maximum of four steps. The relatively small differences among these four trees prompted Ferris *et al.* (1981a) to suggest an unresolved trichotomy for human, chimpanzee and gorilla. Brown *et al.* (1982) and Hixson and Brown (1986) sequenced the 896 bp fragment containing the ND–4 and ND–5 genes and the 12S rRNA gene, respectively. Again, the results were equivocal in that neither study demonstrated strong support, in terms of internode length, for a *Homo/ Pan* clade.

A recent study by Ruvolo *et al.* (1991) on nucleotide sequence variation in the cytochrome c oxidase subunit II gene (COII) provided quite a different result. These data revealed a very large internode supporting the *Homo/Pan* clade. If one considers these recent mitochondrial sequence data plus past evidence from certain other molecular studies, a *Homo/Pan* relationship is supported. Given this conclusion, an explanation as to why there is so much variation with respect to the amount of molecular change supporting a *Homo/Pan* clade remains to be provided, and this is good justification for a continuation of comparative molecular studies on this group in the future.

Conclusion. As indicated by the examples of comparative studies on both Old World monkeys and hominoid primates, the mtDNA molecule has considerable potential as a source for phylogenetic comparisons of recently diverged populations and species as well as older taxa. This conclusion is based on the following facts. (1) The mitochondrial genome is composed of protein-coding, RNA-coding, and non-coding sequences, all of which differ with respect to their overall rates of evolution. This differential in rates of evolution can allow for detailed studies of populations such as seen in the recent comparisons of D-loop variation in African human populations (Vigilant *et al.*, 1989, 1991). If the correct gene is chosen (e.g. the COII gene for hominoid primates: Ruvolo *et al.*, 1991), higher level relationships can be explored. (2) Even within a coding sequence, several different levels of divergence can be analysed separately. For instance, amino acid replacements and transversions associated with primarily first and second position changes in proteins are more likely to be linear with time and less subject to saturation effects (Miyamoto and Boyle, 1989). Therefore, more comparisons involving more divergent taxa can include those regions of the genome less likely to demonstrate extreme homoplasy. (3) Finally, restriction site and fragment comparisons have proven to be very useful for studies of geographic variation and population structure in primates. The ease with which many of these analyses can be done makes this approach very useful for studies of other primate species.

Dating evolutionary events

One of the most critical applications of molecular data has been the dating of some crucial evolutionary events. Tracing our evolutionary history to a single ancestral 'Mother Eve' (Johnson *et al.*, 1983; Cann *et al.*, 1987) and assigning dates to branching events in the human/chimp/ gorilla clade (Templeton, 1981; Brown *et al.*, 1982; Hixson and Brown, 1986) are just two of the issues that have been addressed by examination of the mitochondrial genome. These studies, as well as more recent work on other higher primates (Harihara *et al.*, 1988a; Hayasaka *et al.*, 1986, 1988a,b; Zhang and Shi, 1989), are predicated on the existence of a clock-like rate of molecular change (Melnick, 1990). For example, if different nucleotide bases were substituted for 1% of the bases in the mitochondrial genome of a population every million years, then two populations that have been separated from each other for one million years would differ at 2% of their nucleotide positions. A species–species comparison that reveals 3% sequence divergence must reflect 1.5 million years of evolutionary separation.

We are not as optimistic about the use of mtDNA for dating past evolutionary events as we were for population genetic analysis and phylogenetic reconstruction. The molecular clock hypothesis, originally proposed by Zuckerkandl and Pauling (1962), rests on two critical assumptions about the dynamics of mtDNA evolution. First, the amount of genetic variation within a taxon is negligible compared to the divergence between taxa. It follows from this assumption that, at the time of reproductive isolation, the molecular differences between 'parent' and 'daughter' species are essentially non-existent. Secondly, the rate at which mtDNA mutations have accumulated must be well documented (or 'calibrated') using independent estimates of divergence, and this rate must be approximately the same in all populations or species (so-called rate constancy).

Intraspecific variation. It is clear from the data on Old World monkeys (Table 6.2) that intraspecific variation is *not* inconsequential. Among the cercopithecines (i.e. macaques), sequence divergence between populations of the same species ranges from 0.2 to 8.4%. Among the hominoids, intraspecific sequence divergence falls roughly in the monkey range (see above), with the highest values found between orangutan populations separated by water. Given the ubiquity of large interpopulational differences within a species, it is quite possible that at the time of separation between two incipient species the levels of divergence are already quite high ($d = 0.02$–0.03). If this is true, these initial differences must be accounted for in order to get an accurate estimate of the differences that have accumulated since separation. The only way to estimate the differences at time zero is to use extant intraspecific variation. Far too little attention has been paid to this potential problem and many clock estimates include significant errors related to an incorrect assumption of intraspecific homogeneity.

Calibration and rate constancy.   Rates of molecular evolution have played an essential role in the development of the molecular clock and the estimation of primate divergence times. Different rates of amino acid and nucleotide sequence evolution in nuclear genes are well documented (King and Jukes, 1969; Dickerson, 1971; Britten, 1986), and considerable debate persists as to the cause and accurate documentation of rate heterogeneity (Sarich, 1972; Sarich and Cronin, 1977; Easteal, 1985; Wu and Li, 1985; Li and Wu, 1987).

Our research on macaque mtDNA is an example of a study that casts doubt on the critical assumption of mtDNA evolutionary rate constancy. For instance, the level of mtDNA difference between four populations of pig-tailed macaque, *Macaca nemestrina*, categorized into two subspecies, has revealed unusually high rates of sequence divergence (A. K. Williams *et al.*, unpublished data). In this study each population is separated from the others by a water barrier, and the dates of the last land-bridge connections are well known using palaeobathymetric data. Thus, we can estimate the rate of sequence divergence without reliance on fossil dates which may be problematic.

Comparing the macaque populations on the Malaysian Peninsula and Sumatra, we find a sequence difference of 1.0%. If we apply the accepted molecular clock divergence rate of 2% per million years, which was empirically derived based on hominoid data (Brown *et al.*, 1979), a separation time of 500 000 years is indicated. However, the bathymetric data suggest that the last separation of these populations occurred only 18 000 years ago. Hence, the differences in mtDNA genomes of these two populations are about 30 times greater than would be expected based on the standard molecular clock.

A similar story emerges when one examines differences between the Mentawai island populations of Siberut and Pagai on the one hand and

Sumatra on the other. Here, sequence divergences range from 4.4 to 5.3%, indicating 2.2–2.7 million years of separation, while the bathymetric estimate is an order of magnitude less, about 180 000 years ago. In fact, the fossil record suggests that there were no species of macaque monkeys in Southeast Asia 2.0 million years ago.

Mitochondrial DNA sequence divergence between eastern and western populations of *Macaca mulatta* is also much higher than would be expected from geological evidence of population separation (D. J. Melnick *et al.*, unpublished data). The boundary between the two major *M. mulatta* haplotypes, which differ by over 4%, coincides with the Brahmaputra River valley in Bangladesh. A large glacier moved down this valley about 180 000 years ago, that apparently isolated these two *M. mulatta* populations. The geographic separation of the two mtDNA haplotypes has been maintained, despite restoration of nuclear gene flow resulting from male migration, because female macaques tend to be highly philopatric. Nevertheless, the two *M. mulatta* haplotypes should have diverged by much less than 1% if their molecular clock ticked at a constant 2% per million years.

The macaque data demonstrate that the molecular clock based on other mammals must be re-calibrated if it is to be applied to macaques. However, the question remains, why the degree of mtDNA sequence divergence is so high between closely related macaque populations or species. There are two possible explanations. (1) The mtDNA haplotypes observed began to diverge long before the two populations became isolated. Thus, they would have coexisted in a single population or species prior to the split, and the sorting of maternal lineages subsequent to the initial isolation led to the absence of each haplotype in one of the two daughter populations or species. This is clearly a possibility given the high levels of geographically structured, intraspecific mtDNA diversity found in many macaque species. (2) It is also conceivable, however, that the molecular clock simply runs faster in macaques than in other mammals.

Another way to address the issue of molecular clock rates involves the use of computer-simulated populations to examine the dynamics of sequence evolution (G. A. Hoelzer *et al.*, unpublished data). The simulated populations consist of 25 local social groups distributed in a 5×5 geographical grid. Females in each group exist in a four-stage dominance hierarchy, and dominance rank is passed from mother to daughter, as in macaques. The effects of population size, individual migration rate, group fission rate, geographic constraints on dispersal, and the association between dominance rank and fitness, can all be explored in the model. Using this approach we found that a reduction in the rate of nucleotide base substitutions occurred when high-ranking females had much greater fitness than females of lower rank. Despite this influence over substitution rate, the average rate at which genetic differences accumulated between individuals from different local groups was not affected by any variables in the model. In fact, the average rate of sequence divergence obtained by simulation closely paralleled the conventionally used rate of 2% per million years. However, the stochastic variance in this rate was very high. Hence, there is no reason to believe that rates of molecular change are constant over time, even in the same population, nor are estimates of time based on the average rate of divergence likely to be reliable.

Dating the origin of modern Homo sapiens.   The origin of and relationships among modern populations of *Homo sapiens* have been a major issue in the anthropological community for many years, and currently there are two contrasting viewpoints as to origin. First, Howells (1976) proposed the 'Noah's Ark' hypothesis, that modern populations of *H. sapiens* descended from a single population of modern humans, and that these modern humans quickly displaced all existing populations. Secondly, Wolpoff *et al.* (1984) proposed a 'multi-regional hypothesis' (a modified version of an earlier one by Coon, 1962), whereby modern humans arose simultaneously across the world. This hypothesis does not advocate a unique recent ancestry but rather suggests that modern humans arose in different regions from a polytypic species (*Homo erectus*), followed by gene flow between regions.

Interpretations of mtDNA variation within and among populations of *Homo sapiens* by at least some authors (Stoneking *et al.*, 1986; Cann *et al.*, 1987) suggest that modern humans originated within approximately the last 150 000–300 000 years, descending from a common female ancestor (or population) in Africa. Therefore, this interpretation supports the hypothesis proposed by Howells (1976) by suggesting a single origin for modern humans from a recent common ancestor. The problem with the interpretations of the mtDNA data is that the patterns of mtDNA variation observed and the evidence from the fossil record are equivocal.

What is the evidence supporting the 'Noah's Ark or Garden of Eden Hypothesis'? Cann *et al.* (1987) examined 147 individual humans from five geographic populations using 'high resolution' restriction site mapping of mtDNA. This analysis showed the 134 different mtDNA haplotypes which revealed a considerable amount of admixture with respect to geographic locality. The primary conclusion as to the African origin and time estimates was based on the observation that a phylogenetic analysis of the 134 haplotypes revealed seven divergent African haplotypes (grouping separately from the other 127 haplotypes) that effectively define the placement of the midpoint root of the tree. This phylogenetic analysis led Cann *et al.* (1987) to suggest that the most ancient haplotypes are African, and a comparison of levels of divergence of these haplotypes was used to calculate time since divergence under the assumption that mtDNA incorporates mutations in a clock-like fashion. On the assumption of a rate of mtDNA divergence of 2–4% per million years, the time since divergence from a common ancestor was calculated.

How accurate is the mtDNA molecular clock? Because Cann *et al.* (1987) did not actually calibrate the rate of mtDNA evolution, the assumption of a 2–4% per million rate of divergence was considered to be tentative. Therefore, Stoneking *et al.* (1986) attempted to provide a calibration using intraspecific comparisons of human populations with the following assumptions. (1) Human populations that colonized a specific region at a particular time provide an important calibration point if these populations have remained isolated. (2) Migration with respect to these isolated human populations is considered to be unidirectional, thus there should be no exchange between the recently founded populations and the original colonizing population. (3) The mtDNA haplotypes found in the isolated region should reflect the original types from the primary colonization. These authors used aboriginal populations from Papua New Guinea, Australia, and America to estimate the rate of mtDNA divergence subsequent to founding from a colonizing source such as Asia. The divergence rate calculated from these comparisons ranged from 1.8 to 9.3% per million years, a range suggested by the investigators as supporting the calibrations and dates proposed by Cann *et al.* (1987).

A more recent study by Vigilant *et al.* (1989) investigated nucleotide sequence variation in the hypervariable parts of the human mitochondrial D-loop. A total of 16 individuals from the !Kung and Herero peoples from southern Africa, 37 Pygmy samples, seven samples from Nigeria, eight Asians, one aboriginal Australian, four Europeans, and eight African Americans were examined for D-loop variation, and a genealogical tree was constructed for the individual haplotypes. The chimpanzee (*Pan troglodytes*) was taken as the outsider group for these comparisons, and the node connecting chimpanzee and humans was used to estimate the divergence rate for the D-loop. The percentage divergence seen between chimpanzee and human was estimated to be 13.6% (42% corrected for multiple hits), and using a chimpanzee/human time since divergence estimate of 5 ($\pm$2) million years provided an average rate of

8.4% per million years for control rēgion divergence. This rate estimate suggests that the common ancestor of modern human mtDNAs existed in Africa approximately 238 000 years ago, a value within the range suggested by Cann *et al.* (1987).

The analysis of variation in the human D-loop region has recently been extended to 189 people of diverse geographic origin, including 121 Africans (Vigilant *et al.*, 1991). This analysis appears to lend additional support to a recent (150 000–250 000 years ago), African origin of modern *Homo sapiens*. However, several researchers (Templeton, 1992; Hedges *et al.*, 1992) have reanalysed the data of Vigilant *et al.* and have demonstrated that alternative phylogenetic trees with African and non-African haplotypes at the root are as parsimonious as those containing a purely African root. Indeed, the existing data do not seem sufficient to resolve the question of geographic origin.

The mtDNA data and interpretations of those data by Cann *et al.* (1987) have revived debate as to whether or not modern humans arose in Africa and dispersed outward, displacing indigenous inhabitants. Some investigators have vehemently opposed the Cann *et al.* (1987) hypothesis for several reasons. First, some palaeoanthropologists (Wolpoff *et al.*, 1984; Wolpoff, 1988) argue that the existing fossil evidence supports a continuity between archaic and modern human populations, an observation contrary to the Cann *et al.* (1987) hypothesis. Other palaeontologists argue that the earliest *Homo sapiens* fossils occur in Africa dating between 100 000–125 000 years before present (Rightmire, 1984; Stringer and Andrews, 1988). Secondly, an Asian origin for *Homo sapiens* has been proposed based on the interpretations of mtDNA data produced from another molecular laboratory (Blanc *et al.*, 1983; Johnson *et al.*, 1983), and a reanalysis of the D-loop data does not conclusively support an African root for the human mtDNA tree. If one considers the facts that the most ancient mtDNA haplotypes in the human sample produced by Cann *et al.* (1987) are African and that African mtDNA haplotypes are more genetically divergent from one another, then the argument for an African origin becomes the most parsimonious. Thirdly, as indicated for the Old World monkey patterns of mtDNA and nuclear gene variation, a consideration of population structure and other demographic features of primate populations must be considered when interpreting population-level phylogenies, especially mtDNA phylogenies. Increased population subdivision, population bottlenecks, and the mode of lineage sorting associated with mtDNA variation may affect the rates and patterns of mtDNA divergence. If human populations in Africa have been more subdivided with minimum dispersal between populations, the increased level of divergence seen in African populations may be a reflection of structure rather than antiquity of mtDNA types. However, this interpretation of the current genetic data does require a more complex model. Finally, the time of origin of modern human populations based on mtDNA divergence is the most questionable. The error associated with these estimates is large as a result of the uncertainty of particular calibration points, assumptions used to estimate a divergence rate, and the amount of stochasticity associated with such small time estimates. If arguments as to the time of origin of modern humans rely on precise dates spanning no more than a few hundred thousand years in either direction, the mtDNA evidence must be considered with some reservations.

Conclusion. The use of mtDNA to date past evolutionary events should be taken as a very crude method. Both general assumptions underlying the application of an mtDNA molecular clock are suspect. First, the divergence between any two mtDNA haplotypes, representing different populations or different species, does not necessarily reflect the time since population isolation, because the two haplotypes may have coexisted in the same population for a long period prior to isolation. Indeed, intraspecific diversity in primate mtDNA sequences can be substantial. Secondly, empirical data suggest that mtDNA molecular clocks may not run at the same rates for different taxa. This point is also indicated by a computer simulation model, which further suggests that time estimates based on mtDNA sequence differences are quite unreliable, even when the molecular clock has been properly calibrated.

Summary

This review briefly describes what is known of the primate mitochondrial genome, how the genome can be examined for variation between individuals, and the population genetic and phylogenetic data that have already been collected. This material can be summarized as follows:

1. mtDNA, because it is small, clonally inherited through the matriline without recombination, and rapidly evolving, is a source of easily interpretable molecular data for detailed evolutionary studies both within and between primate taxa.
2. The methods of molecular biology have allowed a complete description of the nucleotide sequence in human mtDNA, as well as sequences of portions of the mtDNA genome of a number of other primates. Restriction enzyme fragment and site data have been collected on an even larger number of primate species.

3. The population genetics of mtDNA are influenced by the maternal inheritance of the molecule and a general tendency in primates toward female philopatry. These factors may explain why in most higher primates (including aboriginal humans) mtDNA variation exhibits a greater degree of geographic structure than does nuclear gene variation. The differences between the two genomes severely limit the inferences one can draw from mtDNA data about the overall genetic affinities of populations.

4. In contrast to the complications encountered in population genetic analysis, reconstructing biogeographical events and phylogenetic relationships, the mtDNA genome is an exceptionally good source of information. Much of its value as a phylogenetic tool derives from the variable rates of nucleotide substitution in different parts of the genome, allowing for the reconstruction of higher order taxonomic relationships using slowly evolving sequences and species and population phylogenies using the more rapidly evolving parts. Despite these advantages one must be aware that mtDNA gene trees are not necessarily the same as species trees, especially for clades of recent origin, because extant mtDNA haplotypes within one species may predate one or more species events in the clade.

5. It is doubtful if estimates of mtDNA sequence divergence can be used to give very accurate measures of the age of a cladogenic split. Large intraspecific differences, temporal fluctuations in substitution rate, and the lack of a universal rate across clades seriously complicate the conversion of molecular divergence into time. We suggest that while a time estimate can be obtained it should be viewed as a rather crude measure with a very large error term around it.

We have obviously raised a number of cautionary notes about interpreting mtDNA data. Nevertheless, on balance, this molecule can be exceptionally useful in analysing the evolutionary process at the population level or higher. With the advent of many new molecular techniques, such a PCR, the use of mtDNA to address ever more complex evolutionary questions of anthropological interest will undoubtedly increase.

References

Adkins, R. M. and Honeycutt, R. L. (1991). Molecular phylogeny of the superorder Archonta. *Proceedings of the National Academy of Sciences USA*, **88**, 10317–21.

Allard, M. W. (1990). Ribosomal gene evolution and the resolution of rodent phylogeny. PhD Dissertation, Harvard University.

Allard, M. W., Ellsworth, D. L. and Honeycutt, R. L. (1991). The production of single stranded DNA suitable for sequencing using the polymerase chain reaction. *BioTechniques*, **10**, 24–6.

Allard, M. W. and Honeycutt, R. L. (1992). Nucleotide sequence variation in the mitochondrial 12S rDNA gene and the phylogeny of African mole rats (Rodentia: Bathyergidae). *Molecular Biology and Evolution*, **9**, 27–40.

Anderson, S., Bankier, A. T., Barrell, B. G., Bruijn, M. H. L., Coulson, A. R., Drouin, J., Eperon, I. C., Nierlich, D. P., Roe, B. A., Sanger, F., Schreier, P. H., Smith, A. J. H., Staden, R. and Young, I. G. (1981). Sequence and organization of the human mitochondrial genome. *Nature*, **209**, 457–65.

Anderson, S., Debruijn, M. H. L., Coulson, A. R., Eperon, I., Sanger, F. and Young, I. G. (1982). Complete sequence of bovine mitochondrial DNA: Conserved features of the mammalian mitochondrial genome. *Journal of Molecular Biology*, **156**, 683–717.

Andrews, P. (1987). Aspects of hominoid phylogeny. In *Molecules and Morphology in Evolution: Conflict or Compromise?*, ed. C. Patterson, pp. 21–53. Cambridge: Cambridge University Press.

Aquadro, C. F. and Greenberg, B. D. (1983). Human mitochondrial DNA variation and evolution: analysis of nucleotide sequences from seven individuals. *Genetics*, **103**, 287–312.

Arnason, U., Gullberg, A. and Widegren, B. (1991). The complete nucleotide sequence of the mitochrondial DNA of the fin whale, *Balaenoptera physalas*. *Journal of Molecular Evolution*, **33**, 556–8.

Attardi, G. (1985). Animal mitochondrial DNA: an extreme example of genetic economy. *International Review of Cytology*, **93**, 93–145.

Avise, J. C. (1986). Mitochondrial DNA and the evolutionary genetics of higher animals. *Philosophical Transactions of the Royal Society, London B*, **312**, 325–42.

Avise, J. C., Arnold, J., Ball, R. M., Bermingham, E., Lamb, T., Neigel, J. E., Reeb, C. A. and Saunders, N. C. (1987). Intraspecific phylogeny: the mitochondrial bridge between population genetics and systematics. *Annual Review of Ecology and Systematics*, **18**, 489–522.

Avise, J. C., Neigel, J. E. and Arnold, J. (1984). Demographic influences on mitochondrial DNA lineage survivorship in animal populations. *Journal of Molecular Evolution*, **20**, 99–105.

Barrell, B. G., Bankier, A. T. and Drouin, J. (1979). A different genetic code in human mitochondria. *Nature*, **282**, 189–94.

Bennet, D., Alexander, L. J., Crozer, R. H. and MacKinlay, A. G. (1988). Are megabats flying primates? Contrary evidence from a mitochondrial DNA sequence. *Australian Journal of Biological Science*, **4**, 327–32.

Bentzen, P., Leggett, W. C. and Brown, G. G. (1988). Length and restriction site heteroplasmy in the mitochondrial DNA of American shad (*Alosa sapidissima*). *Genetics*, **118**, 509–18.

Bermingham, E., Lamb, T. and Avise, J. C. (1986). Size polymorphism and heteroplasmy in the mitochondrial DNA of lower vertebrates. *Journal of Heredity*, **77**, 249–52.

Beverley, S. M. and Wilson, A. C. (1984). Molecular evolution of *Drosophila* and the higher Diptera. II. A time scale for fly evolution. *Journal of Molecular Evolution*, **21**, 1–13.

Bibb, M. J., Van Etten, R. A., Wright, C. T., Walberg, M. W. and Clayton, D. A. (1981). Sequence and gene organization of mouse mitochondrial DNA. *Cell*, **26**, 167–80.

Blanc, H., Chen, K. H., D'Amore, M. and Wallace, D. C. (1983). Amino acid change associated with the major polymorphic Hinc II site of Oriental and Caucasian mitochondrial DNAs. *American Journal of Human Genetics*, **235**, 167–76.

Boursot, P., Yonekawa, H. and Bonhomme, F. (1987). Heteroplasmy in mice with a deletion of a large coding region of mitochondrial DNA. *Molecular Biology and Evolution*, **4**, 46–55.

Brega, A., Gardella, R., Semino, O., Morpurgo, G., Astaldi-Ricotti, F. B., Wallace, D. C. and Santachiara-Benerecetti, A. S. (1986a). Genetic studies on the Tharu population of Nepal: restriction endonuclease polymorphisms of mitochondrial DNA. *American Journal of Human Genetics*, **39**, 502–12.

Brega, A., Scozzari, R., Maccioni, L., Iodice, C., Wallace, D. C., Bianco, I., Cao, C. and Santachiara-Benerecetti, A. S. (1986b). Mitochondrial DNA polymorphisms in Italy, I. Population data from Sardinia and Rome. *Annals of Human Genetics*, **50**, 327–38.

Britten, R. J. (1986). Rates of DNA sequence evolution differ between taxonomic groups. *Science*, **231**, 1393–8.

Brown, W. M. (1980). Polymorphism in mitochondrial DNA of humans as revealed by restriction endonuclease analysis. *Proceedings of the National Academy of Sciences USA*, **77**, 3605–9.

(1981). Mechanisms of evolution in animal mitochondrial DNA. *Annals of the New York Academy of Science*, **361**, 119–34.

(1983). Evolution of animal mitochondrial DNA. In *Evolution of Genes and Proteins*, ed. M. Nei and R. K. Koehn, pp. 62–88. Sunderland, MA: Sinauer Press.

(1985). The mitochondrial genome of animals. In *Molecular Evolutionary Genetics*, ed. R.J. MacIntyre, pp. 95–130. New York: Plenum Press.

Brown, W. M., George, M. and Wilson, A. C. (1979). Rapid evolution of animal mitochondrial DNA. *Proceedings of the National Academy of Sciences USA*, **76**, 1967–71.

Brown, W. M., Prager, E. M., Wang, A. and Wilson, A. C. (1982). Mitochondrial DNA sequences of primates: tempo and mode of evolution. *Journal of Molecular Evolution*, **18**, 225–39.

Brown, G. G. and Simpson, M. V. (1982). Novel features of animal mtDNA evolution as shown by sequences of two rat cytochrome oxidase subunit II genes. *Proceedings of the National Academy of Sciences USA*, **79**, 3246–50.

Caccone, A. and Powell, J. R. (1989). DNA divergence among hominoids. *Evolution*, **43**, 925–42.

Cann, R. L., Brown, W. M. and Wilson, A. C. (1984). Polymorphic sites and the mechanism of evolution in human mitochondrial DNA. *Genetics*, **106**, 479–99.

Cann, R. L., Stoneking, M. and Wilson, A. C. (1987). Mitochondrial DNA and human evolution. *Nature*, **325**, 31–6.

Cann, R. L. and Wilson, A. C. (1983). Length mutations in human mitochondrial DNA. *Genetics*, **104**, 699–711.

Cantatore, P., De Benedetto, C., Fadaleta, G., Gallerani, R., Kroon, A. M., Holtrop, M., Lanave, C., Pepe, G., Quagliariello, C., Saccone, C. and Sbisa, E. (1982). The nucleotide sequences of several tRNA genes from rat mitochondria: common features and relatedness to homologous species. *Nucleic Acids Research*, **10**, 3279–89.

Case, J. T. and Wallace, D. C. (1981). Maternal inheritance of mitochondrial DNA polymorphisms in cultured human fibroblasts. *Somatic Cell Genetics*, **7**, 103–8.

Chomyn, A., Cleeter, W. J., Ragan, C. I., Riley, M., Doolittle, R. F. and Attardi, G. (1986). URF6, last unidentified reading frame of human mtDNA, codes for a NADH dehydrogenase subunit. *Science*, **234**, 614–18.

Chomyn, A., Mariottine, P., Cleeter, M. W. J., Ragan, C. I., Marsuno-Yagi, A., Hatefi, Y., Doolittle, R. F. and Attardi, G. (1985). Six unidentified reading frames of human mitochondrial DNA encode components of the respiratory chain NADH dehydrogenase. *Nature*, **314**, 592–7.

Clary, D. O., Goddard, J. M., Martin, S. C., Fauron, C. M. R. and Wolstenholme, D. R. (1982). *Drosophila* mitochondrial DNA: a novel gene order. *Nucleic Acids Research*, **10**, 6619–37.

Clayton, D. A. (1982). Replication of animal mitochondrial DNA. *Cell*, **28**, 693–705.

(1984). Transcription of the mammalian mitochondrial genome. *Annual Review of Biochemistry*, **53**, 573–94.

Clutton-Brock, T. H. (1989). Female transfer and inbreeding avoidance in social mammals. *Nature*, **337**, 70–2.

Coon, C. S. (1962). *The Origin of Races*. New York: Knopf.

Dawid, I. B. and Blackler, A. W. (1972). Maternal and cytoplasmic inheritance of mitochondrial DNA in *Xenopus*. *Developmental Biology*, **29**, 152–61.

DeFrancesco, L., Attardi, G. and Croce, G. M. (1980). Uniparental propagation of mitochondrial DNA in mouse–human cell hybrids. *Proceedings of the National Academy of Sciences USA*, **77**, 4079–83.

Delson, E. (1980). Fossil macaques, phyletic relationships, and a scenario of deployment. In *The Macaques: Studies in Ecology, Behavior, and Evolution*, ed. D. G. Lindberg, pp. 10–30. New York: Van Nostrand-Reinhold.

Denaro, M., Blanc, H., Johnson, M. J., Chen, K. W., Wilmsen, E., Cavalli-Sforza, L. L. and Wallace, D. C. (1981). Ethnic variation in Hpa I endonuclease cleavage patterns of human mitochondrial DNA. *Proceedings of the National Academy of Sciences USA*, **78**, 5768–72.

Densmore, L. D., Brown, W. M. and Wright, J. W. (1985). Length variation and heteroplasmy are frequent in mitochondrial DNA from parthenogenetic and bisexual lizards (genus *Cnemodophorus*). *Genetics*, **110**, 689–707.

Desjardins, P. and Morais, R. (1990). Sequence and gene organization of the chicken mitochondrial genome: a novel gene order in higher vertebrates. *Journal of Molecular Biology*, **212**, 599–634.

Dickerson, R. E. (1971). The structure of cytochrome c and the rates of molecular evolution. *Journal of Molecular Evolution*, **1**, 26–45.

Disotell, T. R., Honeycutt, R. L. and Ruvulo, M. (1992). Mitochrondial DNA phylogeny of the old-world monkey tribe Papionini. *Molecular Biology and Evolution*, **9**, 1–13.

Djian, P. and Green, H. (1989). Vectorial expansion of the involucrin gene and the relatedness of the hominoids. *Proceedings of the National Academy of Sciences USA*, **86**, 8447–51.

Dowling, T. E., Moritz, C. and Palmer, J. D. (1990). Nucleic Acids II: Restriction site analysis. In *Molecular Systematics*, ed. D. Hillis and C. Moritz, pp. 250–317. Sunderland, MA: Sinauer Press.

Easteal, S. (1985). Generation time and the rate of molecular evolution. *Molecular Biology and Evolution*, **2**, 450–3.

Engels, W. R. (1981). Estimating genetic divergence and genetic variability with restriction endonucleases. *Proceedings of the National Academy of Sciences USA*, **78**, 6329–33.

Farris, J. S. (1970). Methods for computing Wagner trees. *Systematic Zoology*, **34**, 21–34.

(1988). HENNIG86. Department of Ecology and Evolution. SUNY, Stoneybrook, New York.

Felsenstein, J. (1982). Numerical methods for inferring evolutionary trees. *Quarterly Review of Biology*, **57**, 379–404.

(1988). Phylogenies from molecular sequences: inference and reliability. *Annual Review of Genetics*, **22**, 521–65.

(1989). PHYLIP: Phylogenetic Inference Package, Version 3.3. Department of Genetics. University of Washington, Seattle.

Ferris, S. D., Wilson, A. C. and Brown, W. M. (1981a). Evolutionary tree for apes and humans based on cleavage maps of mitochondrial DNA. *Proceedings of the National Academy of Sciences USA*, **78**, 2432–6.

Ferris, S. D., Brown, W. M., Davidson, W. S. and Wilson, A. C. (1981b). Extensive polymorphism in the mitochondrial DNA of apes. *Proceedings of the National Academy of Sciences USA*, **78**, 6319–23.

Ferris, S. D., Sage, R. D., Prager, E. M., Ritte, U. and Wilson, A. C. (1983). Mitochondrial DNA evolution in mice. *Genetics*, **105**, 681–721.

Fitch, W. M. and Margoliash, E. (1967). Construction of phylogenetic trees. *Science*, **155**, 279–84.

Fleagle, J. G. (1988). *Primate Adaptation and Evolution*. San Diego: Academic Press.

Fooden, J. (1980). Classification and distribution of living macaques. In *The Macaques: Studies in Ecology, Behavior, and Evolution*, ed. D.G. Lindberg, pp. 1–9. New York: Van Nostrand-Reinhold.

(1990). The bear macaque, *Macaca arctoides*: a systematic review. *Journal of Human Evolution*, **19**, 607–86.

Fooden, J. and Lanyon, S. M. (1989). Blood protein allele frequencies and phylogenetic relationships in Macaca: a review. *American Journal of Primatology*, **17**, 209–41.

Foran, D. R., Hixson, J. E. and Brown, W. M. (1988). Comparison of ape and human sequences that regulate mitochondrial DNA transcription and D-loop DNA synthesis. *Nucleic Acids Research*, **16**, 5841–61.

Gadaleta, G., Pepe, G., DeCandia, G., Quagliariello, C., Sbisa, E. and Saccone, C. (1989). The complete nucleotide sequence of the *Rattus norvegicus* mitochondrial genome: cryptic signals revealed by comparative analysis between vertebrates. *Journal of Molecular Evolution*, **28**, 497–516.

George, M., Jr (1982). Mitochondrial DNA evolution in Old World monkeys. PhD Dissertation, University of California, Berkeley.

Gibbons, A. (1990). Our chimp cousins get that much closer. *Science*, **250**, 376.

Giles, R. E., Blanc, H., Cann, H. M. and Wallace, D. C. (1980). Maternal inheritance of human mitochondrial DNA. *Proceedings of the National Academy of Sciences USA*, **77**, 6715–19.

Gillespie, J. H. (1986). Variability of evolutionary rates of DNA. *Genetics*, **113**, 1077–91.

Gingerich, P. D. (1986). Temporal scaling of molecular evolution in primates and other mammals. *Molecular Biology and Evolution*, **3**, 205–21.

Gojobori, T., Ishii, K. and Nei, M. (1982a). Estimation of average number of nucleotide substitutions when the rate of substitution varies with nucleotide. *Journal of Molecular Evolution*, **18**, 414–23.

Gojobori, T., Li, W. H. and Graur, D. (1982b). Patterns of nucleotide substitution in pseudogenes and functional genes. *Journal of Molecular Evolution*, **18**, 360–9.

Goodman, M. (1976). Toward a genealogical description of the primates. In *Molecular Anthropology*, ed. M. Goodman and R. E. Tashian, pp. 321–53. New York: Plenum Press.

Goodman, M., Braunitzer, G., Stangl, A. and Schrank, B. (1983). Evidence on human origins from haemoglobins of African apes. *Nature*, **303**, 346–8.

Gortz, G. and Feldman, H. (1982). Nucleotide sequence of the cytochrome b gene and adjacent regions from rat liver mitochondrial DNA. *Current Genetics*, **5**, 221–5.

Gotoh, O., Hayashi, J.-I., Yonekawa, H. and Tagashira, Y. (1979). An improved method for estimating sequence divergence between related DNAs, from changes in restriction endonuclease cleavage sites. *Journal of Molecular Evolution*, **14**, 301.

Grosskopf, R. and Feldman, H. (1981). Analysis of a DNA segment from rat liver mitochondrial DNA containing the genes from cytochrome oxidase subunits I, II, and III, ATPase subunit 6, and several tRNA genes. *Current Genetics*, **4**, 151–8.

Groves, C. P. (1980). Speciation in *Macaca*: the view from Sulawesi. In *The Macaques: Studies in Ecology, Behavior, and Evolution*, ed. D. G. Lindberg, pp. 84–124. New York: Van Nostrand-Reinhold.

Gyllensten, U., Wharton, D. and Wilson, A. C. (1985). Maternal inheritance of mitochondrial DNA during backcrossing of two species of mice. *Journal of Heredity*, **76**, 321–4.

Hale, L. R. and Singh, R. S. (1986). Extensive size variation and heteroplasmy in mitochondrial DNA among geographic populations of *D. melano-*

226   D. J. Melnick, G. A. Hoelzer and R. L. Honeycutt

gaster. *Proceedings of the National Academy of Sciences USA*, **83**, 8813–17.

Harihara, S., Saitou, N., Hirai, M., Aoto, N., Tero, K., Cho, F., Honjo, S. and Omoto, K. (1988a). Differentiation of mitochondrial DNA types in *Macaca fascicularis*. *Primates*, **29**, 117–27.

Harihara, S., Saitou, N., Hirai, M., Gojobori, T., Park, K. S., Misawa, S., Ellepola, S. B., Ishida, T. and Omoto, K. (1988b). Mitochondrial DNA polymorphism in Japanese living on Hokkaido. *Japanese Journal of Human Genetics*, **31**, 134–43.

Hauswirth, W. W., Van De Walle, M. J., Laipis, P. J. and Olivio, P. D. (1984). Heterogeneous mitochondrial DNA D-loop sequences in bovine tissue. *Cell*, **37**, 1001–7.

Hayasaka, K., Gojobori, T. and Horai, S. (1988a). Molecular phylogeny and evolution of primate mitochondrial DNA. *Molecular Biology and Evolution*, **5**, 626–44.

Hayasaka, K., Horai, S., Gojobori, T., Shotake, T., Nowaza, K. and Matsunaga, E. (1988b). Phylogenetic relationships among Japanese, rhesus, Formosan and crab-eating monkeys, inferred from restriction enzyme analysis of mitochondrial DNAs. *Molecular Biology and Evolution*, **5**, 270–81.

Hayasaka, K., Horai, S., Shotake, T., Nozawa, K. and Matsunaga, E. (1986). Mitochondrial DNA polymorphism in Japanese monkeys, *Macaca fuscata*. *Japanese Journal of Genetics*, **61**, 345–59.

Hayashi, J.-I., Tagashira, Y. and Yoshida, M. C. (1985). Absence of extensive recombination between inter- and intraspecies mitochondrial DNA in mammalian cells. *Experimental Cell Research*, **160**, 387–95.

Hedges, S. B., Kumar, S., Tamura, K. and Stoneking, M. (1992). Human origins and analysis of mitochondrial DNA sequences. *Science*, **255**, 737–9.

Hennig, W. (1966). *Phylogenetic Systematics*. Urbana, IL: University of Illinois Press.

Higuchi, R., Von Beroldingen, C. H., Sensabaugh, G. F. and Erlich, H. A. (1988). DNA typing from single hairs. *Nature*, **332**, 543–6.

Higuchi, R., Bowman, B., Freiberger, M., Ryder, O. A. and Wilson, A. C. (1984). DNA sequences from the quagga, an extinct member of the horse family. *Nature*, **312**, 282–4.

Hillis, D. M. and Moritz, C. (1990). *Molecular Systematics*. Sunderland, MA: Sinauer Press.

Hixson, J. E. and Brown, W. M. (1986). A comparison of the small ribosomal RNA genes from the mitochondrial DNA of the great apes and humans: sequence, structure, evolution, and phylogenetic implications. *Molecular Biology and Evolution*, **3**, 1–18.

Hoelzer, G. A. and Melnick, D. J. (1992). Genetic and evolutionary relationships of the macaques. In *Evolutionary Ecology and Behavior of the Macaques*, ed. J. Fa and D. G. Lindlung. Cambridge: Cambridge University Press (in press).

Honeycutt, R. L., Nelson, K., Schlitter, D. A. and Sherman, P. W. (1991). Genetic variation within and between populations of the naked mole-rat: evidence from nuclear and mitochondrial genes. In *Biology of the Naked*

Mole-Rat, ed. P. W. Sherman, R. Alexander and J. U. M. Jarvis, pp. 195–208. Princeton, NJ: Princeton University Press.

Honeycutt, R. L. and Wheeler, W. C. (1989). Mitochondrial DNA: variation in human and higher primates. In *DNA Systematics: Human and Higher Primates*, ed. S. K. Dutta and W. Winter, pp. 91–129. Boca Raton, FL: CRC Press.

Horai, S., Gojobori, T. and Matsunaga, E. (1984). Mitochondrial DNA polymorphism in Japanese. I. Analysis with restriction enzymes of six base pair recognition. *Human Genetics*, **68**, 324–32.

Horai, S. and Matsunaga, E. (1986). Mitochondrial DNA polymorphism in Japanese. II. Analysis with restriction enzymes of four or five base pair recognition. *Human Genetics*, **72**, 105–17.

Howells, W. W. (1976). Explaining modern man: evolutionists versus migrationists. *Journal of Human Evolution*, **5**, 477–96.

Hutchison, C. A., Newbold, C. E., Potter, S. S. and Edgell, M. H. (1974). Maternal inheritance of mammalian mitochondrial DNA. *Nature*, **251**, 536–8.

Innis, M. A., Gelfand, D. H., Sninsky, J. J. and White, T. J. (1990). *PCR Protocols: A Guide to Methods and Applications*. San Diego, CA: Academic Press.

Irwin, D. M., Kocher, T. D. and Wilson, A. C. (1991). Evolution of the cytochrome b gene of mammals. *Journal of Molecular Evolution*, **32**, 128–44.

Johnson, M. J., Wallace, D. C., Ferris, A. C., Rattazi, M. C. and Cavalli-Sforza, L. L. (1983). Radiation of human mitochondrial DNA types analyzed by restriction endonuclease cleavage pattern. *Journal of Molecular Evolution*, **19**, 255–71.

Jukes, T. H. and Cantor, C. R. (1969). Evolution of protein molecules. In *Mammalian Protein Metabolism*, ed. H. N. Munro, pp. 21–32. New York: Academic Press.

Kaplan, N. (1983). Statistical analysis of restriction enzyme map data and nucleotide sequence data. In *Statistical Analysis of DNA Sequence Data*, ed. B. S. Weir, pp. 75–106. New York: Marcel Dekker.

Kawamoto, Y., Takenaka, O. and Botoisworo, E. (1982). Preliminary report on genetic variations within and between species of Sulawesi macaques. *Kyoto University Overseas Research Reports, Studies in Asian Non-human Primates*, **2**, 23–37.

Kimura, M. (1980). A simple method for estimating evolutionary rate of base substitutions through comparative studies of nucleotide sequences. *Journal of Molecular Evolution*, **16**, 111–20.

(1983). *The Neutral Theory of Molecular Evolution*. Cambridge: Cambridge University Press.

King, J. L. and Jukes, T. H. (1969). Non-Darwinian evolution. *Science*, **164**, 788–98.

Kobayashi, M., Seki, T., Yaginuma, K. and Koike, K. (1981). Nucleotide sequences of small ribosomal RNA and adjacent transfer RNA genes in rat mitochondrial DNA. *Gene*, **16**, 272–307.

Koike, K., Kobayashi, M., Yaginuma, K., Taira, M., Yoshida, E. and Imai, M. (1981). Nucleotide sequence and evolution of the rat mitochondrial cytochrome b gene containing the ochre termination codon. *Gene*, **20**, 177–85.

Koop, B., Goodman, M., Xu, P., Chan, K. and Slightom, J. L. (1986). Primate η-globin DNA sequences and man's place among the great apes. *Nature*, **319**, 234–8.

Langley, C. H. and Fitch, W. M. (1974). An examination of the constancy of the rate of molecular evolution. *Journal of Molecular Evolution*, **3**, 161–77.

Lansman, R. A., Avise, J. C. and Huettel, M. D. (1983). Critical experimental test of the possibility of "paternal leakage" of mitochondrial DNA. *Proceedings of the National Academy of Sciences USA*, **80**, 1969–71.

Lansman, R. A., Shade, R. O., Shapira, J. F. and Avise, J. C. (1981). The use of restriction endonucleases to measure mitochondrial DNA sequence relatedness in natural populations. III. Techniques and potential applications. *Journal of Molecular Evolution*, **17**, 214–26.

Lewontin, R. C. (1972). The apportionment of human diversity. *Evolutionary Biology*, **6**, 381.

Li, W.-H. and Wu, C.-I. (1987). Rates of nucleotide substitution are evidently higher in rodents than in man. *Molecular Biology and Evolution*, **4**, 74–82.

Maddison, W. P. and Maddison, D. R. (1992). MacClade 3.0. Sunderland, MA: Sinauer Associates.

Maniatis, T., Fristch, E. F. and Sambrook, J. (1989). *Molecular Cloning: A Laboratory Manual*, 2nd edn. Cold Spring Harbor: Cold Spring Harbor Publications.

Margulis, L. (1981). *Symbiosis in Cell Evolution*. San Francisco: W. H. Freeman.

Melnick, D. J. (1987). The genetic consequences of primate social organization: A review of macaques, baboons, and vervet monkeys. *Genetica*, **73**, 117–35.

(1988). The genetic structure of a primate species: rhesus monkeys and other cercopithecine primates. *International Journal of Primatology*, **9**, 195–231.

(1990). Molecules, evolution and time. *TREE*, **5**, 772–3.

Melnick, D. J. and Hoelzer, G. A. (1992). Differences in male and female macaque dispersal lead to contrasting distributions of nuclear and mitochondrial DNA variation. *International Journal of Primatology* (in press).

Melnick, D. J., Jolly, C. J. and Kidd, K. K. (1984a). The genetics of a wild population of rhesus monkeys (*Macaca mulatta*) I. Genetic variability within and between social groups. *American Journal of Physical Anthropology*, **63**, 341–60.

Melnick, D. J. and Kidd, K. K. (1985). Genetic and evolutionary relationships among Asian macaques. *International Journal of Primatology*, **6**, 123–60.

Melnick, D. J., Pearl, M. C. and Richard, A. F. (1984b). Male migration and inbreeding in wild rhesus monkeys. *American Journal of Primatology*, **7**, 229–43.

Miyamoto, M. M. and Boyle, S. M. (1989). The potential importance of mitochondrial DNA sequence data to eutherian mammal phylogeny. In *The Hierarchy of Life*, ed. B. Fernholm, K. Bremer and H. Jornvall, pp. 437–50. Amsterdam: Elsevier Science Publications.

Miyamoto, M. M. and Cracraft, J. (1991). *Phylogenetic Analysis of DNA Sequences.* New York: Oxford University Press.

Miyamoto, M. M., Kraus, F. and Ryder, O. A. (1990). Phylogeny and evolution of antlered deer determined from mitochondrial DNA sequences. *Proceedings of the National Academy of Sciences USA*, **87**, 6127–31.

Miyamoto, M. M., Slightom, J. L. and Goodman, M. (1987). Phylogenetic relations of humans and African apes from DNA sequences in the $\psi\eta$-globin region. *Science*, **238**, 369–73.

Miyamoto, M. M., Tanhauser, S. M. and Laipis, P. J. (1989). Systematic relationships in the artiodactyl tribe Bovini (family Bovidae), as determined from mitochondrial DNA sequences. *Systematic Zoology*, **38**, 342–9.

Miyata, T. and Yasunaga, T. (1980). Molecular evolution of mRNA: a method for estimating evolutionary rates of synonymous and amino acid substitutions from homologous nucleotide sequences and its application. *Journal of Molecular Evolution*, **16**, 23–36.

Monnat, R. J. and Loeb, L. A. (1985). Nucleotide sequence preservation of human mitochondrial DNA. *Proceedings of the National Academy of Sciences USA*, **82**, 2895–9.

Moore, J. (1984). Female transfer in primates. *International Journal of Primatology*, **5**, 537–89.

Moritz, C. (1987). Tandem duplications in animal mitochondrial DNAs: variation in incidence and gene content among lizards. *Proceedings of the National Academy of Sciences USA*, **84**, 7183–7.

Moritz, C. and Brown, W. M. (1986). Tandem duplication of D-loop and ribosomal RNA sequences in lizard mitochondrial DNA. *Science*, **233**, 1425–7.

Moritz, C., Dowling, T. E. and Brown, W. M. (1987). Evolution of animal mitochondrial DNA: relevance for population biology and systematics. *Annual Review of Ecology and Systematics*, **18**, 269–92.

Napier, J. R. and Napier, P. H. (1967). *A Handbook of Living Primates.* London: Academic Press.

Nei, M. (1973). Analysis of gene diversity in subdivided populations. *Proceedings of the National Academy of Sciences USA*, **70**, 3321–3.

(1982). Evolution of human races at the gene level. In *Human Genetics, Part A: The Unfolding Genome*, ed. B. Bonne-Tamir, P. Cohen and R. N. Goodman, pp. 167–81. New York: Alan R. Liss.

(1985). Human evolution at the molecular level. In *Population Genetics and Molecular Evolution*, ed. K. Aoki and T. Ohta, pp. 41–64. Tokyo: Japan Science Society Press.

(1987). *Molecular Evolutionary Genetics.* New York: Columbia University Press.

Nei, M. and Li, W. H. (1979). Mathematical model for studying genetic variation in terms of restriction endonucleases. *Proceedings of the National Academy of Sciences USA*, **76**, 5269–73.

Nei, M. and Tajima, F. (1981). DNA polymorphism detectable by restriction endonucleases. *Genetics*, **97**, 145–63.

(1983). Maximum likelihood estimation of the number of nucleotide substitutions from restriction sites data. *Genetics*, **105**, 207–17.

Nowaza, K., Shotake, T., Kawamoto, Y. and Tanabe, Y. (1982). Population genetics of Japanese monkeys. II. Blood protein and polymorphisms and population structure. *Primates*, **23**, 252–71.

Pääbo, S. (1989). Ancient DNA: extraction, characterization, molecular cloning and enzymatic amplification. *Proceedings of the National Academy of Sciences USA*, **86**, 1939–43.

Pääbo, S., Gifford, J. A. and Wilson, A. C. (1988). Mitochondrial DNA sequences from a 7000 year old brain. *Nucleic Acids Research*, **16**, 9775–87.

Pusey, A. E. and Packer, C. (1987). Dispersal and Philopatry. In *Primate Societies*, ed. B. B. Smuts, D. L. Cheney, R. M. Seyfarth, R. W. Wrangham and T. T. Strusaker, pp. 250–66. Chicago: University of Chicago Press.

Ramharack, R. and Deeley, R. G. (1987). Structure and evolution of primate cytochrome c oxidase subunit II gene. *Journal of Biological Chemistry*, **262**, 14014–21.

Rand, D. M. and Harrison, R. G. (1986). Mitochondrial DNA transmission genetics in crickets. *Genetics*, **114**, 955–70.

Reilly, J. G. and Thomas, C. A. (1980). Length polymorphisms, restriction site variation, and maternal inheritance of mitochondrial DNA of *Drosophila melanogaster*. *Plasmid*, **3**, 109–15.

Roe, B. A., Ma, D. P., Wilson, R. K. and Wong, J. F. H. (1985). The complete nucleotide sequence of the *Xenopus laevis* mitochondrial genome. *Journal of Biological Chemistry*, **260**, 9759–74.

Ruvolo, M., Disotell, T. R., Allard, M. W., Brown, W. M. and Honeycutt, R. L. (1991). Resolution of the African hominoid trichotomy by use of a mitochondrial gene sequence. *Proceedings of the National Academy of Sciences USA*, **88**, 1570–4.

Sade, D. S. (1972). A longitudinal study of social behavior of rhesus monkeys. In *The Functional and Evolutionary Biology of Primates*, ed. R. H. Tuttle, pp. 378–98. Chicago, IL: Aldine Press.

Saiki, R. K., Scharf, S., Faloona, F., Mullis, K. B., Horn, G. T., Erlich, H. A. and Arnheim, N. (1985). Enzymatic amplification of beta-globin genomic sequences and restriction site analysis for diagnosis of sickle cell anemia. *Science*, **230**, 1350–4.

Saiki, R. K., Bugawan, T. L., Horn, G. T., Mullis, K. B. and Erlich, H. A. (1986). Analysis of enzymatically amplified beta-globin and HLA-DQ alpha DNA with allele specific oligonucleotide probes. *Nature*, **324**, 163–6.

Saiki, R. K., Gelfand, D. H., Stoffel, S., Scharf, S. J., Higuchi, R., Horn, G. T., Mullis, K. B. and Erlich, H. A. (1988). Primer-directed enzymatic amplification of DNA with a thermostable DNA polymerase. *Science*, **239**, 487–91.

Saitou, N. and Nei, M. (1987). The neighbor-joining method: a new method for reconstructing phylogenetic trees. *Molecular Biology and Evolution*, **4**, 406–25.

Sanger, F., Nicklen, S. and Coulson, A. R. (1977). DNA sequencing with chain-terminating inhibitors. *Proceedings of the National Academy of Sciences USA*, **74**, 5463–7.

Santachiara-Benerecetti, A. S., Scozzari, R., Semino, O., Torroni, A., Brega, A. and Wallace, D. C. (1988). Mitochondrial DNA polymorphism in four Sardinian villages. *Annals of Human Genetics*, **52**, 327–40.

Sarich, V. M. (1972). Generation time and albumin evolution. *Biochemical Genetics*, **7**, 205–12.

Sarich, V. M. and Cronin, J. E. (1977). Generation length and rate of hominoid molecular evolution. *Nature*, **269**, 354–5.

Sarich, V. M., Schmid, C. W. and Marks, J. (1989). DNA hybridization as a guide to phylogenies: a critical analysis. *Cladistics*, **5**, 3–32.

Sarich, V. M. and Wilson, A. C. (1967). Immunological time scale for hominid evolution. *Science*, **158**, 1200–3.

Schurr, T. G., Ballinger, S. W., Gan, Y. Y., Hodge, J. A., Merriwether, D. A., Lawrence, D. N., Knowler, W. C., Weiss, K. M. and Wallace, D. C. (1990). Amerindian mitochondrial DNAs have rare Asian mutation at high frequencies, suggesting they derived from four primary maternal lineages. *American Journal of Human Genetics*, **46**, 613–23.

Shotake, T. (1981). Population genetic study of natural hybridization between *Papio anubis* and *P. hamadryas*. *Primates*, **20**, 285–308.

Sibley, C. G. and Ahlquist, J. E. (1984). The phylogeny of the hominoid primates, as indicated by DNA–DNA hybridization. *Journal of Molecular Evolution*, **20**, 2–15.

(1987). DNA hybridization evidence of hominoid phylogeny: results from an expanded data set. *Journal of Molecular Evolution*, **26**, 99–121.

Sibley, C. G., Comstock, J. A. and Ahlquist, J. E. (1990). DNA hybridization evidence of hominoid phylogeny: a reanalysis of the data. *Journal of Molecular Evolution*, **30**, 202–36.

Smouse, P. E., Speilman, R. S. and Park, M. H. (1982). Multiple-locus allocation of individuals to groups as a function of the genetic variation within and differences among human populations. *American Naturalist*, **119**, 445–61.

Sneath, P. H. A. and Sokal, R. R. (1973). *Numerical Taxonomy*. San Francisco, CA: W. H. Freeman.

Solignac, M., Monnerot, M. and Mounolou, J.-C. (1983). Mitochondrial DNA heteroplasmy in *Drosophila mauritiana*. *Proceedings of the National Academy of Sciences USA*, **80**, 6942–6.

Solus, J. F. and Eisenstadt, J. M. (1984). Retention of mitochondrial DNA species in somatic cell hybrids using antibiotic selection. *Experimental Cell Research*, **151**, 299–305.

Southern, S. O., Southern, P. J. and Dizon, A. E. (1988). Molecular characterization of a cloned dolphin mitochondrial genome. *Journal of Molecular Evolution*, **28**, 32–42.

Spuhler, J. N. (1988). Evolution of Mitochondrial DNA in monkeys, apes, and humans. *Yearbook of Physical Anthropology*, **31**, 15–48.

Stoneking, M., Bhatia, K. and Wilson, A. C. (1986). Rate of sequence divergence estimated from restriction maps of mitochondrial DNAs from Papua New Guinea. *Cold Spring Harbor Symposia on Quantitative Biology*, **51**, 433–9.

232   D. J. Melnick, G. A. Hoelzer and R. L. Honeycutt

Stoneking, M., Jorde, L. B., Bhatia, K. and Wilson, A. C. (1990). Geographic variation in human mitochondrial DNA from Papua New Guinea. *Genetics*, **124**, 717–33.

Stringer, C. B. and Andrews, P. (1988). Genetic and fossil evidence for the origin of modern humans. *Science*, **239**, 1263–8.

Swofford, D. L. (1990). Phylogenetic Analysis Using Parsimony (PAUP) Version 3.0. Champaign, IL: Illinois Natural History Survey.

Swofford, D. L. and Olsen, G. J. (1990). Phylogeny Reconstruction. In *Molecular Systematics*, ed. D. M. Hillis and C. Moritz, pp. 411–501. Sunderland, MA: Sinauer Press.

Szalay, F. S. and Delson, E. (1979). *Evolutionary History of the Primates*. New York: Academic Press.

Tajima, F. and Nei, M. (1984). Estimation of evolutionary distance between nucleotide sequence. *Molecular Biology and Evolution*, **1**, 269–85.

Takahata, N. (1989). Gene genealogy in three related populations: consistency probability between gene and population trees. *Genetics*, **122**, 959–68.

Tanhauser, S. M. (1985). Evolution of mitochondrial DNA: patterns and rate of change. PhD Dissertation, University of Florida, Gainesville, FL.

Templeton, A. R. (1981). Phylogenetic inference from restriction endonuclease cleavage site maps with particular reference to the evolution of human apes. *Evolution*, **37**, 221–44.

Templeton, A. R. (1992). Human origins and analysis of mitochondrial DNA sequences. *Science*, **255**, 737.

Thomas, W. K., Pääbo, S., Villablanca, F. X. and Wilson, A. C. (1990). Spatial and temporal continuity of kangaroo rat populations shown by sequencing mitochondrial DNA from museum specimens. *Journal of Molecular Evolution*, **31**, 101–12.

Turner, T. (1981). Blood protein variation in a population of Ethiopian vervet monkeys (*Cercopithecus aethiops aethiops*). *American Journal of Physical Anthropology*, **55**, 225–32.

Upholt, W. B. (1977). Estimation of DNA sequence divergence from comparison of restriction endonuclease digests. *Nucleic Acids Research*, **10**, 2225–40.

Vigilant, L., Pennington, R., Harpending, H., Kocher, T. D. and Wilson, A. C. (1989). Mitochondrial DNA sequences in single hairs from a southern African population. *Proceedings of the National Academy of Sciences USA*, **86**, 9350–4.

Vigilant, L., Stoneking, M., Harpending, M., Hawkes, K. and Wilson, A. C. (1991). African populations and the evolution of human mitochondrial DNA. *Science*, **253**, 1503–7.

Wallace, D. C., Garrison, K. and Knowler, W. C. (1985). Dramatic founder effects in Amerindian mitochondrial DNAs. *American Journal of Physical Anthropology*, **68**, 149–55.

Wheeler, W. C. (1990). Combinatorial weights in phylogenetic analysis: a statistical parsimony procedure. *Cladistics*, **6**, 269–75.

Whittam, T. S., Clark, A. G., Stoneking, M., Cann, R. L. and Wilson, A. C. (1986). Allelic variation in human mitochondrial genes based on patterns of restriction site polymorphism. *Proceedings of the National Academy of Sciences USA*, **83**, 9611–15.

Williams, A. K. (1990). The evolution of mitochondrial DNA in the silenus-sylvanus species group of macaques. PhD Dissertation, Columbia University, New York.

Williams, P. L. and Fitch, W. M. (1989). Finding the minimal change in a given tree. In *The Hierarchy of Life*, ed. B. Fernholm, K. Bremer and H. Jornvall, pp. 453–70. Amsterdam: Elsevier Science Publications.

Wilson, A. C., Cann, R. L., Carr, S. M., George, M., Gyllensten, U. B., Helm-Bychowski, K. M., Higuchi, R. F., Palumbi, S. R., Prager, E. M., Sage, R. D. and Stoneking, M. (1985). Mitochondrial DNA and two perspectives on evolutionary genetics. *Biological Journal of the Linnean Society*, **26**, 375–400.

Wolpoff, M. H. (1988). Multi-regional evolution: the fossil alternative to Eden. In *The Human Revolution: Behavioural and Biological Perspective in the Origins of Modern Humans*, ed. P. Mellars and C. B. Stringer, pp. 62–108. Princeton, NJ: Princeton University Press.

Wolpoff, M. H., Xinzhi, W. and Thorne, A. G. (1984). Modern *Homo sapiens* origins: a general theory of hominid evolution involving the fossil evidence from east Asia. In *The Origins of Modern Humans: A World Survey of the Fossil Evidence*, ed. F. H. Smith and F. Spencer, pp. 441–83. New York: Alan R. Liss.

Wolstenholme, D. R., Fauron, C. M.-R. and Goddard, J. M. (1982). Nucleotide sequence of *Rattus norvegicus* mitochondrial DNA that includes the genes for tRNA^{ile}, tRNA^{gln}, and tRNA$^{f\text{-}met}$. *Gene*, **20**, 63–9.

Wolstenholme, D. R. and Clary, D. O. (1985). Sequence evolution of *Drosophila* mitochondrial DNA. *Genetics*, **109**, 725–44.

Wrischnik, L. A., Higuchi, R. G., Stoneking, M., Erlich, H. A., Arnheim, N. and Wilson, A. C. (1987). Length mutations in human mitochondrial DNA: direct sequencing of enzymatically amplified DNA. *Nucleic Acids Research*, **15**, 529–42.

Wu, C.-I. and Li, W. H. (1985). Evidence for higher rates of nucleotide substitution in rodents than in man. *Proceedings of the National Academy of Sciences USA*, **82**, 1741–5.

Yang, W. and Zhou, X. (1988). rRNA genes are located far away from the D-loop region in Peking duck mitochondrial DNA. *Current Genetics*, **13**, 351–5.

Zhang, Y. and Shi, X. (1989). Mitochondrial DNA polymorphism in five species of the genus *Macaca*. *Chinese Journal of Genetics*, **16**, 326–38.

Zuckerkandl, E. and Pauling, L. (1962). Molecular disease, evolution and genic heterogeneity. In *Horizons in Biochemistry*, ed. M. Kasha and B. Pullman, pp. 189–225. New York: Academic Press.

7 Beads and string: the genome in evolutionary theory

JONATHAN MARKS

Introduction

One of the most pervasive images in biology over the last hundred years arrived with the union of the chromosome theory of heredity with Mendelian genetics, shortly after the turn of the century. With the knowledge that the visible chromosomes carried the hereditary information, and that the genes carried the hereditary information, it was a reasonable inference that the genes lay on or in the chromosomes. Heredity was soon recognized (largely through the work of Thomas Hunt Morgan and his students) to be governed by elementary particles, each of which had a particular localization on one of relatively few scalar filaments. The chromosomes, those filaments, were as a 'string of beads', in a simile apparently first used by Castle (1924). The beads themselves – the interesting part of the necklace – were the genes, and who would buy a necklace that had a yard of string but just a few beads?

Perhaps the largest theoretical retrenchment in evolutionary genetics over the last few decades has been the recognition that the genome – the DNA matrix in which genes are embedded – is a far more extensive and complex entity than was formerly thought (Bodmer, 1981). In other words, the beads may be less important than the string, or, perhaps more appropriately, the beads are just a special kind of string. This argument comes from two fundamental inferences in molecular genetics. First, there is a quantitative DNA paradox: protein-coding information, translated by the rules of the genetic code, decoded in the 1960s, accounts for perhaps 1% of the DNA in a cell. Secondly, there is a dynamic component to the genome, a battery of mutational processes whose causes and effects are still poorly understood.

The fact that they are poorly understood, however, does not mean they are insignificant. It is generally held that approximately 10% of the genome consists of genes, on the average, and approximately 10% of an average gene codes for a protein product. Thus, evolution in terms of protein coding sequences accounts for only about 1% of the data.

234

Reasoning from the genes to the genome – the specific to the general – is, in effect, the tail wagging the dog.

The genome is generally defined as the entire DNA complement of a haploid cell, which conceals the fact that there is structure to it, even if the structure is not well understood (Kao, 1985). The genome is not only the context in which genes exist, it is the genes themselves, and their parts. Genes contain large untranslated regions before, after and within the coding sequence; though why such lengths of seemingly useless DNA are needed can only be guessed at (Doolittle, 1987).

One effect of ignoring the context in which genes are encountered has been the reification of the gene: some biologists have given it properties, such as 'selfishness' (Dawkins, 1976). In fact, the attribution of basic properties like replication to genes is just such a fallacy, for the units of DNA replication in the genome ('replicons') are different from, and not apparently correlated with, the units of RNA transcription (i.e. 'genes').

The term 'gene' itself loses much of its traditional meaning under the weight of contemporary molecular analysis. A generation ago, a gene was simultaneously a unit of transcription, a unit of translation into protein information, and a discrete sequence of DNA. Now we know that most of what is transcribed is not translated, so that there is genome *within* genes; further, the boundaries of genes are not clear. Is a promoter (such as CCAAT and TATA) part of a gene or outside it? Is genomic DNA that serves as a binding site, but is not itself transcribed, to be considered a gene? Are genomic repetitive elements genes? Are genes whose RNAs can be alternatively spliced two genes or one? Are coding sequences within non-coding sequences of larger genes two genes or one?

The successful integration of molecular genetics into evolutionary theory will have to involve reasoning from the general to the specific, in this case, from the genome to the genes. The goal of this chapter, therefore, is to discuss the significance of the genome for biological anthropology; to suggest inverting a traditional mode of thought in evolutionary theory, from a focus on the evolution of genes, with the rest of the DNA as 'hitch-hiking', to a focus on the evolution of the *genome*, with the genes as special cases.

Beyond the synthetic theory

The issues to be addressed in this chapter can be brought more sharply into focus by examining the relationships shown in Fig. 7.1. The theory of evolution is an attempt to explain how species come to differ from one another, how organisms come to differ from one another, and how cells

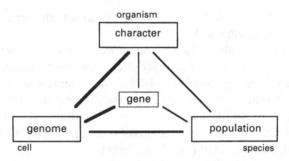

Fig. 7.1. Relationships among genome, character, and population (each representing a different level in a hierarchy: cf. Vrba and Eldredge, 1984), with the gene holding a central place in contemporary evolutionary theory. The genome's relationships to the other entities remain to be explored.

come to differ from one another. Central to this explanation is the gene, which is part of a cell's genome, a determinant of an organism's phenotypes or characters, and which establish specific mate-recognition systems and reproductive incompatibilities among populations.

Darwin's work was explicit in elucidating the relationship between characters and populations: phenotypic characters, in so far as they may be adaptive, are the result of the action of natural selection. The gene, a 20th century concept, has a central place in the structure of modern evolutionary theory. A grasp of its dynamics in populations is the legacy of the synthesis of evolution and population genetics, achieved in the 1930s by Haldane, Wright, and Fisher.

The relationship between gene and character remains a central question in genetics, and one for which evolutionary theory has maintained a long-standing oversimplification: that a gene determines a trait or character. This has been criticized from time to time, most notably by Mayr (1959), who labelled this attitude 'beanbag genetics'. Much of our understanding of gene and character stems from the study of pathologies: isolating a mutation, learning what is wrong with the organism possessing it, and then inferring the gene's role from what is lacking in the organism lacking the gene. The problem, of course, is the pervasive pleiotropy and epistasis we encounter in physiological genetic systems. Thus, although victims of Lesch–Nyhan syndrome possess a mutation of the X chromosome in a specific gene, and they have an uncontrollable urge to bite off their lips and fingertips, one would be hard pressed to argue that the normal allele functions to prevent biting off one's lips. In fact, it codes for

an enzyme called HGPRT (hypoxanthine guanine phosphoribosyl trans-
ferase), an enzyme in purine metabolism; Lesch–Nyhan syndrome is a
pleiotropic effect (Stout and Carey, 1988).

On the other hand, epistasis is so pervasive that while genes affect
characters, one is equally hard pressed to name a single character in a
eukaryotic organism whose genetic basis is understood. This would
require isolating many genes, understanding their regulation and function, and tracking the character ontogenetically as different genes pass in
and out of transcriptional activity. It would also require understanding
how non-genetic factors can influence the appearance of the phenotype
regardless of genotype.

The major unsolved, indeed often unrecognized, questions in evolutionary biology stem from the relations of the genome to the individual
gene, to the evolving population, and to the phenotypic character. In
other words; how is the genome structured, how does genetic evolution
really work, and where do phenotypes ultimately come from? If an
answer cannot be framed here, it may be possible at least to see where the
answers may reside, and how we will recognize them when we find them.

The relationship of genome to gene

Britten and Kohne (1968) were able to dichotomize the genome into
unique-sequence and repetitive components, which varied in proportion
across taxa. The discovery of the redundant nature of much of the
genome had major implications for evolutionary theory. Initially these
were falsely dichotomized and oversimplified: the repetitive DNA caused
morphological change, while the 'molecular clock' serum protein genes
did not (Britten and Davidson, 1971). Rather, the more productive
implications of this discovery concern the ways in which the genome
grows and generates genetic diversity, in addition to the nucleotide
substitutions inherent in molecular evolutionary theory since the
Watson–Crick DNA model was proposed in 1953.

Satellite DNA

One significant component of the genome was first isolated because it
consisted of classes of homogeneous sequences of densities slightly
different from the bulk of the DNA in the rest of the genome. Thus, when
the density of genomic DNA was graphed, this class of DNA appeared as
'satellite' peaks, set apart from the bulk of the genome. Though its

proportion may vary widely among closely related taxa, there seems little doubt that satellite DNA is ubiquitous, certainly among vertebrates. By the prevailing logic of molecular genetics, it is therefore presumably important as well. Surprisingly little, however, is known of the function of this DNA.

What sets satellite DNA apart from other genomic components is (1) the large number of copies and (2) the fact that it is localized, rather than dispersed. The localization is presumably what permits the copies to evolve 'in concert' (cf. below).

The most well-characterized human satellite DNA is known as alpha-satellite, about 172 bp in length and located at the centromere of each human chromosome. This class of satellite DNA is actually a family of related sequences; each chromosome apparently has its own character-istic repeated alpha-satellite subunit, not exactly the same as that on other chromosomes, nor on the homologous chromosome in closely related species (Rosenberg et al., 1978; Manuelidis and Wu, 1978; Alexandrov et al., 1988; Wevrick and Willard, 1989; Waye and Willard, 1989).

The presence of tandem arrays of repeated sequences is highly condu-cive to a process that has come to be known as 'concerted evolution' in which serially homologous DNA sequences are 'corrected' against one another, and thus evolve 'in concert'. The tandem arrays of sequences maintain their homogeneity through time, and do not accumulate mu-tations independently of one another; what mutations do occur are either 'corrected' out, or 'corrected' into other serial homologs. This is presum-ably the cause of the chromosome-specificity and species-specificity of each array of alpha-satellite repeats.

Gene families: the globin paradigm

Even the DNAs of the genes and spacers are not *literally* unique sequences, as most genes are now known to be members of families, structurally similar genes that duplicated and diverged from one another in the remote past. Some of these genes are localized into clusters, such as the β-globins and α-globins that code for the haemoglobin subunits (see Chapter 5), but others are more scattered; indeed, the two blood globin clusters are members of a single family, located on chromosomes 11 and 16, respectively.

The implication of gene families is pervasive serial homology in the unique-sequence genome. This means that much unique-sequence DNA is not, broadly speaking, unique. While there may be no other sequences

α-globin gene cluster
Chromosome 16

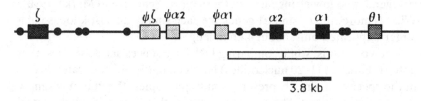

●   *Alu* repeats

▨   paralogous genes and pseudogenes

▭   α-globin duplication block

Fig. 7.2. Structure of the human α-globin gene cluster, a tiny portion of the genome, located near the tip of the short arm of chromosome 16. These 30 000 bases of DNA account for fewer than 500 amino acids of functional protein sequence.

exactly identical to a particular 500 bp stretch, there may well be several that differ by a few per cent. Indeed, a critical variable in many molecular genetic experiments (such as Southern blots) is the *stringency* of the hybridization of a DNA probe to the genome; the conditions above which the probe will hybridize as if it were unique-sequence, and below which it may cross-hybridize with other serially homologous sequences (see Chapter 4).

The α-globin gene cluster is shown in Fig. 7.2. It includes seven known genes, all serially homologous (or in the terms of Fitch (1970), 'paralogous'). The 'rubber-stamp' process of gene duplication leaves three fates for a duplicate gene: (1) it can continue to make the same product, if the resulting phenotype is advantageous, (2) it can accumulate mutations, if the resulting phenotypes are neutral, and make a slightly different product that can be co-opted in a tissue or developmental stage and (3) it can accumulate mutations and shut down. In the third case, the gene is called a pseudogene.

The human α-globin cluster gives evidence of all three of these processes. Zeta (ζ) is the functional embryonic alpha-like globin; pseudo-zeta (ψζ) is a pseudogene, functional in a polymorphism; pseudo-alpha–2 (ψα2) is a pseudogene; pseudo-alpha–1 (ψα1) is a pseudogene; alpha–2 (α2) and alpha–1 (α1) make identical proteins; and theta–1 (θ1) makes an RNA that codes for an uncharacterized protein (Marks, 1989). While none of these is literally identical to any other ($\alpha2$ and $\alpha1$, the most similar pair, differ by two point mutations and an insertion of 7 bp in one non-coding region, and by about 16% past the end of the coding sequence), *all* will cross-hybridize with one another to some extent in medium-stringency and low-stringency Southern blots (Schmid and Marks, 1990). What is more, the $\alpha2$ and $\alpha1$ genes are each simply the last kilobase of a larger duplicated unit, about 4 kb in length.

Often one finds, when comparing DNA sequences across species, that a short sequence (1–20 nucleotides) has been tandemly duplicated in one of the species. If already present in multiple copies, the unit may simply vary in copy number. The process governing this pattern is called 'strand slippage'. For example, when comparing the intergenic DNA between $\alpha2$ and $\alpha1$ across species, a 17 bp sequence in the human was found to be present as two copies, side by side, in the homologous location in the orangutan (J. Marks, J. P. Shaw and C. K. J. Shen, unpublished data). Examples in the published homologous DNA sequences for the hominoids are not hard to find. In the pseudo-eta ($\psi\eta$)-globin DNA sequence, position 1391 begins two tandem sequences of TTCTA in human, chimpanzee, gorilla and orangutan, and (with one point mutation) in baboon; but in the rhesus macaque, three sequences are present. Position 1705 begins a sequence of 23 A's in the human, but only 17 in chimpanzee and gorilla, and 12 A's interrupted by 3 T's in the orangutan (Goodman *et al.*, 1989).

Interspersed elements: the Alu paradigm

Among, and even occasionally within, the genes are estimated to be several hundred thousand *Alu* repeats (Schmid and Jelinek, 1982). These are the best characterized of the short interspersed elements, or SINEs, in the human genome. Unlike satellite DNA, these repetitive elements are not localized, but rather are interspersed throughout the genome. Their significance may be appreciated from the fact that the 5% of the genome that *Alus* alone represent is probably larger than the proportion of the genome actually translated into protein products.

To date, nearly a hundred *Alus* have been sequenced, and they have a characteristic structure. They are dimers of a fundamental 160 bp sequence (probably originally derived from a highly conserved RNA gene whose product is involved in the translational process, called 7SL RNA), with a 30 bp deletion in the first monomer and an A-rich tail. The monomer is similar to a repetitive sequence from rodents and a prosimian (Schmid and Shen, 1985; Deininger and Daniels, 1986), but the *Alu* itself seems to be restricted in taxonomic distribution – a probable haplorhine synapomorphy.

*Alu*s are particularly interesting because of the question that their presence and distribution naturally raises: how did they get there? They have internal promoter regions, similar to the ribosomal and transfer RNA genes that are transcribed by RNA polymerase III (ordinary mRNAs are transcribed by Pol II); and indeed they can be induced to transcribe RNA *in vitro*. *In vivo*, however, there is no good evidence that individual *Alu*s exercise that capability (Paulson and Schmid, 1986), which suggests that they are all products of one or a few 'master' sequences in the genome.

The distribution of *Alu* repeats is the result of a process that is still not well known. A transposable genetic element, or transposon, is a DNA sequence with the capability of moving from place to place in the genome; it can do so either by leaving its original position, or by making a copy which can then insert itself elsewhere (Temin and Engels, 1984). Some transposons, apparently related to viral DNA sequences, encode enzymes that, when translated, enable it to be transcribed into an RNA, to be reverse-transcribed into DNA, and to be re-inserted elsewhere (viral retroposons); others transpose without an RNA intermediate (Finnegan, 1989). While these are efficient duplication mechanisms, they are not particularly prolific, and relatively few copies of each of these transposon sequences are present in the genome, though they may play a significant role in evolutionary processes, such as P elements in *Drosophila* (cf. below).

By contrast, *Alu* repeats are considered to be non-viral retroposons (Weiner *et al.*, 1986). The elusive 'master *Alu*' is a more fecund transcriber – its RNA is capable of being processed, reverse-transcribed, and reinserted in abundance – and is apparently not coding for its own enzymes, or for any proteins at all, unlike the *viral* retroposons. Yet the inert, processed DNA copies are present in very large numbers in the human genome.

The distribution of *Alu*s has an apparently non-random feature. In about half the genome they are scattered at an average of 2.5 kb apart (Schmid *et al.*, 1983). This half appears to be the one in which genes are abundant, and which is not stained in the process of G-banding (Korenberg and Rykowski, 1988). In the other half of the genome, they are more diffusely scattered.

The evolutionary effects of interspersed redundancy are probably twofold, both relating to the mutational process. First, a newly generated repeat sequence could insert itself into a location where it compromises

242   J. Marks

the structural integrity of a gene (e.g. Quattrochi *et al.*, 1986); this would presumably be deleterious in the vast majority of cases. Second, the presence of repeats can promote unequal crossing-over, which in turn generates duplications and deletions, and alters the spatial arrangement of portions of the genome. While this may also most often be deleterious (Lehrman *et al.*, 1987; Rouyer *et al.*, 1987; Stoppa-Lyonnet *et al.*, 1990), it could conceivably be a creative source of diversity as well.

The relationship of genome to species

The genetic processes by which two populations, or even two organisms, may be rendered reproductively incompatible are largely unknown. There are several plausible explanations for specific cases, but the general applicability of any particular genetic mechanism is probably very easy to overestimate.

Chromosomal rearrangements

Chromosomal rearrangements have often been invoked as mechanisms in speciation (e.g. White, 1969, 1978; Arnason, 1972; Marks, 1987). First, the chromosomes of related species often appear to be somewhat different. Secondly, given the rarity of genes relative to the size of a chromosome, one could reasonably expect random breakages of chromosomes to be unlikely to affect the structure or expression of any particular gene. Thirdly, chromosomes are intimately involved in the reproductive process: rearrangements to chromosomes must affect the meiotic pairing which will occur in an F_1 hybrid of parents with different chromosome configurations, presumably reducing the chances of producing an F_2. Yet it has been difficult to establish chromosomal rearrangement as a cause, as opposed to a simple correlate, of speciation. In some groups chromosomes differ widely among closely related species (for example, among gibbons, some cercopithecines and some cebids). Across other species, for example papionins, there is little or no chromosomal difference across species. This implies that if chromosomes provide a genetic mechanism for speciation, they do it only in some groups.

Hybrid dysgenesis

Transposons known as P elements are found in *Drosophila melanogaster*, and have an interesting history. Strains of this species recently collected in the wild are known as P strains, whose genomes contain P elements. Some strains collected in the 1960s are P strains, while others are M strains, lacking the elements. Old laboratory stocks, derived from strains collected before the 1960s, are all M strains. This genomic difference holds up even for populations sampled from the same locality at different times (David and Capy, 1988).

The P elements code for proteins which permit them to be copied and transposed, but they also code for their own repressor. Thus, when in the 'native' state, P elements repress themselves, by virtue of proteins in the cytoplasm that they have produced. If a P female mates with an M male, the sperm do not have P elements, but the egg has P elements and repressors in its cytoplasm; the result is normal reproduction, and perhaps the entry of P elements into a new population. If, however, a P male mates with an M female, the zygotes now have cytoplasm lacking P repressors, and P elements contributed by the sperm. The result is that the P elements transcribe, insert, and wreak general mutational havoc; resulting in sterility, a phenomenon known as hybrid dysgenesis (Snyder and Doolittle, 1988; Finnegan, 1989).

Controversy exists as to the general evolutionary significance of this phenomenon. Regardless of whether this is incipient speciation, true anagenetic speciation, or neither (Rose and Doolittle, 1983; Syvanen, 1984), the genome of *Drosophila melanogaster* has been transformed over the last few decades by these elements.

Polyploidy

The overall genome size of mammals is very conservative, though other higher taxa appear to tolerate significant changes in genome size more readily. One way is through an increase in ploidy, the number of chromosome sets normally present in somatic cells (Sparrow and Nauman, 1976). Polyploidy is a common evolutionary phenomenon in plants (Stebbins, 1966), and less common, but known, in non-mammalian animal taxa.

The human genome (and inferentially, the mammalian genome) is believed to bear the remnants of an archaic genome doubling that occurred perhaps 500 million years ago, the best evidence for which consists of similar clusters of paralogous genes on different chromosomes (Comings, 1972; Tolan *et al.*, 1987; Schughart *et al.*, 1989).

Molecular drive

Dover (1982) reified the diverse genomic evolutionary processes under a common name, 'molecular drive', to contrast the rapid and radical changes to the genome with the nucleotide substitutions assumed in the classic formulations of genetic evolutionary theory. It is not clear whether the processes involved in 'molecular drive' are considered to be involved in the generation of reproductive differences among populations (Dover, 1978; Dover et al., 1982), or in the establishment of morphological novelty (Dover, 1986), or in neither – simply the spread of 'selfish DNA', replicating without organismal effect (Doolittle and Sapienza, 1980; Orgel and Crick, 1980).

'Molecular drive' is thus something very general subsuming many different processes which may have diverse results; yet it serves an important function in drawing attention to the fact that the processes underlying evolutionary genetics are mechanically far more complicated than just the spread of point mutations through time.

The relationship of genome to character

The origin of phenotypes may well be the central problem of genetics. Yet the origin of normal phenotypes in any organism is at present an intractable problem even in a synchronic, clinical sense (i.e. where does this character come from?). In a diachronic, evolutionary sense (i.e. how did the genetic machinery that produced *this* character in *this* species come to produce *that* character in *that* species?), virtually nothing is known.

Since characteristics of higher organisms tend to be polygenic, it stands to reason that the genome would be a central concept in developing a corpus of theory to explain phenotypes. However, this requires explaining one poorly understood entity (a phenotype) by recourse to another poorly understood entity (genome structure). Suffice it to say that individual genes affect characters, but the characters are products of the genome, and the genome is considerably more flexible and responsive than is any individual gene (cf. McClintock, 1977). In order to appreciate what implications we face for the evolution of complex phenotypes, it may be necessary to think in terms of some new analogies. Perhaps the studies of change in other kinds of systems may be of assistance in modelling the evolutionary forces at work here. After all, at the same time that anthropologists were rejecting the atomism inherent in describing changes in cultural units over time, because of the systemic nature of culture, population geneticists, led by Fisher, were adopting the analogous notion of describing biological change as a change in the frequency of alleles through time. This approach has been criticized by Mayr since 1959 as 'beanbag genetics'. An alternative approach, emphasizing the interaction among genes in a biological system, was taken by Sewall Wright (cf. 1931, 1932), but it had little impact because of the daunting mathematics (Provine, 1986). Thus, in general, the view of evolution used by population genetics tended to reflect the 'beanbag' approach, even though it did not accurately reflect the way evolution really works (Haldane, 1960; Michod, 1981). It may well be profitable to look to the study of culture change, where the 'beanbag' approach has long been out of favour, to provide analogies to help illuminate this aspect of biological change.

Implications for phylogenetic inference

The most important implication of the new conception of the genome – lots of string, and few beads – lies in the recognition that the evolutionary processes are far more complex than an earlier generation of evolutionary geneticists had conceived them to be. First, the mutational processes are diverse indeed; secondly, there is a hierarchy of selection implied by the differential ability of DNA segments to replicate themselves within the genome. Some organisms can make copies of themselves more efficiently than other organisms, colonizing a terrestrial environment, and some DNA segments can make copies of themselves more efficiently than other DNA segments, colonizing a cellular environment. The former results in the adaptation of species; the latter results in serial homology within the genome.

The second implication of the new genome concept may be even more radical and far-reaching than the first. The upshot of the *Ramapithecus* controversy in the 1970s, coupled with the revolution in molecular technology, left many evolutionary biologists with the feeling that molecular biology would hold the ultimate answers to phylogeny (cf. Gould, 1985; Diamond, 1988). The two techniques most widely discussed in this regard are DNA hybridization and DNA sequencing; and a consideration of the complex nature of homology produced by genomic processes will show that neither is *or probably could be* an 'ultimate' answer for phylogeny.

DNA hybridization

DNA hybridization, though comparing much of the genome (the least redundant half is typically used), nevertheless runs into the problem of serial homology. The purpose of DNA hybridization is to estimate the amount of sequence divergence across the entire unique-sequence genome, by a mass comparison of one species DNA with another species DNA. This is accomplished by hybridizing the DNA of the two species and observing a reduction in the thermal stability of the DNA duplexes formed, when compared against non-hybrid DNA.

The fact, however, that there is pervasive serial homology interspersed throughout the genome means that any DNA sequence from one species can, in principle, hybridize to its ortholog in the other species, *or to a paralog*. A paralogous hybrid will be poorly bonded, and will be less thermally stable than an orthologous hybrid. Thus, the melting profile of DNA from such an experiment tends to have two main peaks (Fig. 7.3), resulting from the pairing of paralogs (low-temperature) and of orthologs (high-temperature). Since, however, it is the orthologous DNA hybrids that are presumably of interest, one needs either to ignore the paralogous hybrids, or to control for any variation in the extent of formation of paralogous hybrids. Unfortunately, this is rarely done in DNA hybridization, and the widespread use of measurements that incorporate paralogous DNA hybrids ($T_{50}H$ and T_m) makes it difficult to assess evolutionary relationships among taxa established on the basis of those measurements. In other words, two pairs of taxa whose DNA hybrids melt at slightly different T_ms could be differently related to one another, or could simply be expressing different amounts of paralogous hybridization.

This paralogy also helps to set an upper limit to the utility of DNA hybridization. The thermal stability of paralogs is a function of when the *genes* duplicated and began to diverge – possibly hundreds of millions of years ago, regardless of when the *taxa* (and orthologs) actually diverged. As one performs a battery of DNA hybridization experiments using DNA from taxa that are increasingly more distantly related, the paralogs and orthologs behave differently (Fig. 7.4). The orthologs become increasingly more poorly paired, but the paralogs stay the same, because they date the same remote event. Thus, as phylogenetic distance increases, proportionately more of what hybridizes is paralogous, and less is orthologous, for the orthologs, now fairly different from one another, do not compete as efficiently against the very different paralogs. Beyond a reduction of about 15 degrees in thermal stability, the paralogous hybrids cannot be separated from the orthologous hybrids in the melting curve. Thus, closely related species can give spurious results from this technique unless variation in the extent of paralogous pairing is discriminated from the comparison of orthologous DNA; and for distant relatives, this cannot be done. The structure of the genome here limits our capacity to understand the extent and significance of differences between the DNAs of two species *en masse*.

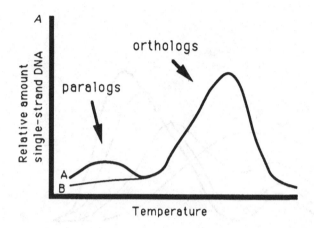

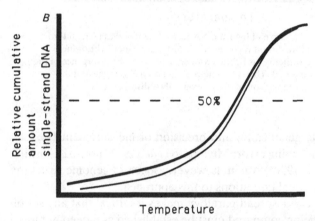

Fig. 7.3. *A*, Typical melting curve of hybrid DNA, in which duplex DNA
molecules dissociate into single strands as the temperature is raised.
Paralogues melt at a lower temperature than orthologues; and the difference
between curves A and B is simply less pairing of paralogues in curve B. *B*, if a
median temperature (50% point) is used to represent 'the melting
temperature' of these hybrids, A and B will have somewhat different values, in
spite of the fact that their orthologous DNAs were equally thermally stable.

DNA sequencing

Where DNA hybridization has the advantage of measuring the divergences of many genes rather than just one, DNA sequencing has the
advantage of permitting the direct comparison of character state trans-
formations across taxa. Obviously all modes of phylogenetic inference
have to transcend the problems of polymorphism (which appears to be a

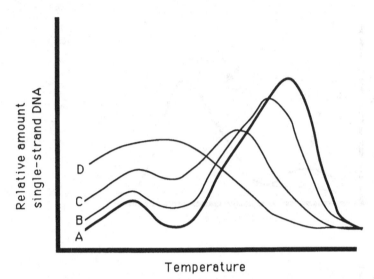

Fig. 7.4. Melting curve of hybrid DNA from close relatives (A), and from successively more distant relatives (B, C, D). Paralogous hybrids increase in proportion to orthologous hybrids with an increase in the divergence of the taxa being compared. This sets an upper limit on the usefulness of the technique as a comparison of orthologous DNA divergence.

significant variable genetically) and precision of measurement (i.e. sequencing and transcribing error: Miyamoto *et al.*, 1987 (note 31); Orkin, 1987; Brunak *et al.*, 1990). Again, however, aspects of genome structure suggest some additional limitations to this approach.

First, the fact of widespread paralogy makes it critical that any set of DNA sequences being compared must be established as indeed homologous (i.e. orthologous). For example, there are about 450 genes for ribosomal RNA divided among the short arms of human chromosomes 13, 14, 15, 21 and 22. These regions exchange genetic information, which generates complex patterns of serial homology among the genes (Arnheim *et al.*, 1980; Krystal *et al.*, 1981). In a study of the rRNA genes of the great apes, in order to obtain phylogenetic information, one would therefore presumably wish to avoid the obvious problem of base sequence heterogeneity across paralogs of the 450 genes, and make certain to study orthologs (Williams *et al.*, 1988). Then, one would still have the process of concerted evolution among serial homologs (and unlinked serial homologs) operating, but its effect upon the phylogenetic comparison would presumably be reduced. Gonzalez *et al.* (1990), however, in attempting such a phylogenetic comparison, used a *paralog* from each

	6864		6883
Human	TTTTT	TTTAA	TTTTAACT
Chimpanzee	TTTTT	AA	TTTTAACT
Gorilla	TTTTT	TTTAAA	TTTTAACT
Orangutan	TTTTT	TTA	TTTTAACT

β-globin region DNA sequence and numbering
from Goodman *et al.* (1989)

Fig. 7.5. Complex homology in a short stretch of DNA sequence.

species. Consequently, we cannot know how reliable a phylogenetic inference based upon them might be for closely related species.

The other problem that accrues is that of inferring homology, and inferring the evolutionary events that occurred to homologous nucleotides, given a relationship of orthology between two DNA sequences from different species. In principle, this may seem relatively straightforward, since each nucleotide has four possible characters (or five, including 'deletion'), and base substitutions can be inferred by simple comparison. But because of small stretches of redundancy, strand slippage can introduce deletions and insertions in similar locations in close relatives. Figure 7.5 shows the difficulty in inferring evolutionary events from orthologous DNA sequences in which strand slippage occurs. The sequence is from Goodman *et al.* (1989), and shows that the human sequence TTTAA is homologous to the chimpanzee sequence AA, the gorilla sequence TTTAAA, and the orangutan sequence TTA. One can, of course, infer a series of evolutionary events to account for this pattern, but the problem is that one can infer *several different* series of evolutionary events, and it is impossible to tell which is right.

Another complicating factor again stems from the fact that the genome is a context in which the gene is embedded. The genome appears to have local compartments of characteristic base composition known as 'isochores' (Bernardi, 1989). Within a particular isochore a gene may not be as free to mutate widely, regardless of phenotypic effects. Thus, if a gene is located in an isochore that is heavily GC-rich, a mutation of C to T in its coding sequence may be unfavourable, regardless of phenotypic effect.

Similarly, certain mutations (transitions: purine to purine and pyrimidine to pyrimidine) appear to be favoured over others (transversions: purine to pyrimidine and vice versa), at least in mitochondrial DNA. Is this true also for nuclear DNA? If such empirical regularities are found, how may they be incorporated into DNA sequence analysis algorithms? Do we lump sequence data from different systems, such as different genes or mitochondrial and nuclear DNA? Do we lump together events inferred on the basis of strand slippage or deletions with those inferred from nucleotide substitutions? Do we lump together DNA segments with different intensities of natural selection acting upon them? Do we accept the most parsimonious tree without question, or designate a baseline of non-parsimony, below which we consider all trees equivalent? As Penny (1989: 305) wrote, '[C]ollecting sequences is the easy part of any study. The hard part is estimating the accuracy of the results.'

The upshot is that comparing DNA sequences to obtain phylogenies has its own set of unresolved methodological questions. These exist for a particular historical reason: that the present models of genetic evolution, developed in the 1930s and 1940s, are models of *gene* evolution. But the genes are embedded in the genome and are minor components of it, and we now know that the genome evolves in a very complex manner. The evolution of genes, the synthetic theory, can be no more than a special case of the evolution of genomes, for which there is no comprehensive theory at present. It should not be a threat to genetics that sometimes a clear phylogenetic interpretation of the molecular data cannot be made; rather, that should be the challenge to the next generation of evolutionary theorists.

The problem in this regard comes with pre-emptive claims that one or another phylogenetic problem has been 'resolved', usually on the basis of particular data, methods of analysis, and assumptions about evolutionary modes, which upon scrutiny emerge to be narrow, flawed, poorly reasoned, or worse. Again, a consideration of the complexities of genome evolution should reinforce the point that any competent evolutionary or systematic study must acknowledge and interpret all the available data bearing on the subject.

Molecular methods and data simply cannot be taken to *supersede* traditional manners of inferring phylogeny at present: rather, they *augment* them. In this sense, molecular data should be taken as testing hypotheses about phylogeny; but it is a tenet of the hypothetico-deductive method that when a hypothesis fails a test, either the hypothesis is falsified, or the test is flawed. Sometimes it is hard to decide between the alternatives.

Conclusions

The synthetic theory of evolution, besides 'hardening' into a rosy adaptationist paradigm in the post-World War II era (Gould, 1980) also erected its genetic models on a conception of the genome now known to be incomplete. There are four principal reasons why the recognition of genome structure and dynamic complexity is important. First, it extends the scope of the evolutionary processes, for genetic variation can originate in many more ways than was previously thought possible. Secondly, some of these processes may facilitate the generation of reproductive incompatibilities between populations, i.e. speciation. Thirdly, it provides a link to the genetics of physiological systems, which are polygenic in nature and regulated in a complex manner; and calls for a retreat from the heuristic simplicity of beanbag genetics. Fourthly, it explains why there can quite easily be a discordance between the phylogeny of a group of species, and the relationships presented on the basis of a genetic study.

This is an intellectually exciting time to be working in the field of molecular evolutionary genetics. As models of genetic evolution increase in sophistication, it will be with the recognition that genome mechanics govern and subsume gene mechanics. The next major step in evolutionary theory will therefore be the elaboration of a genome theory of evolution that will relate the structure and composition of the hereditary material to its functional units, to the characteristics of organisms, and to the boundaries between species.

Acknowledgements
This manuscript was prepared with assistance from NSF BNS–8819047, and a Junior Faculty Fellowship in the Social Sciences from Yale University.

References
Alexandrov, I. A., Mitkevick, S. P. and Yurov, Y. B. (1988). The phylogeny of human chromosome specific alpha satellites. *Chromosoma*, **96**, 443–53.
Arnason, U. (1972). The role of chromosomal rearrangement in mammalian speciation with special reference to Cetacea and Pinnipedia. *Hereditas*, **70**, 113–18.
Arnheim, N., Crystal, M., Schmickel, R., Wilson, G., Ryder, O. and Zimmer, E. (1980). Molecular evidence for genetic exchanges among ribosomal genes on non-homologous chromosomes in man and apes. *Proceedings of the National Academy of Sciences USA*, **77**, 7323–7.
Bernardi, B. (1989). The isochore organization of the human genome. *Annual Review of Genetics*, **23**, 637–61.
Bodmer, W. F. (1981). Gene clusters, genome organization, and complex

phenotypes: When the sequence is known, what will it mean? *American Journal of Human Genetics*, **33**, 664–82.

Britten, R. and Davidson, E. H. (1971). Repetitive and non-repetitive DNA sequences and a speculation on the origins of evolutionary novelty. *Quarterly Review of Biology*, **46**, 111–38.

Britten, R. and Kohne, D. (1968). Repeated sequences in DNA. *Science*, **161**, 529–40.

Brunak, S., Engelbrecht, J. and Knuden, S. (1990). Cleaning up gene databases. *Nature*, **343**, 123.

Castle, W. E. (1924). *Heredity and Eugenics*, 4th edn. Cambridge, MA: Harvard University Press.

Comings, D. (1972). Evidence for ancient tetraploidy and conservation of linkage groups in mammalian chromosomes. *Nature*, **238**, 455–7.

David, J. R. and Capy, P. (1988). Genetic variation of *Drosophila melanogaster* natural populations. *Trends in Genetics*, **4**, 106–11.

Dawkins, R. (1976). *The Selfish Gene*. Oxford: Oxford University Press.

Deininger, P. and Daniels, G. R. (1986). The recent evolution of mammalian repetitive DNA elements. *Trends in Genetics*, **3**, 76–80.

D'Eustachio, P. and Ruddle, F. (1983). Somatic cell genetics and gene families. *Science*, **220**, 919–24.

Diamond, J. (1988). DNA-based phylogenies of the three chimpanzees. *Nature*, **332**, 685–6.

Doolittle, W. L. (1987). The origin and function of intervening sequences in DNA: A review. *American Naturalist*, **130**, 915–28.

Doolittle, W. F. and Sapienza, C. (1980). Selfish genes, the phenotype paradigm, and genome evolution. *Nature*, **284**, 601–3.

Dover, G. (1982). Molecular drive: A cohesive mode of species evolution. *Nature*, **299**, 111–17.

  (1986). Molecular drive in multi-gene families: How biological novelties arise, spread, and are assimilated. *Trends in Genetics*, **2**, 159–65.

  (1988). Evolution of the third kind. *Nature*, **332**, 402.

Dover, G., Strachan, T., Coen, E. S. and Brown, S. D. M. (1982). Molecular drive. *Science*, **218**, 1069.

Finnegan, D. (1989). Eukaryotic transposable elements and genome evolution. *Trends in Genetics*, **5**, 103–7.

Fitch, W. (1970). Distinguishing homologous from analogous proteins. *Systematic Zoology*, **19**, 99–113.

Gonzalez, I., Sylvester, J., Smith, T., Stambolian, D. and Schmickel, R. (1990). Ribosomal RNA gene sequences and hominoid phylogeny. *Molecular Biology and Evolution*, **7**, 203–19.

Goodman, M., Koop, B., Czelusniak, J., Fitch, D., Tagle, D. and Slightom, J. (1989). Molecular phylogeny of the family of apes and humans. *Genome*, **31**, 316–35.

Gould, S. J. (1980). G. G. Simpson, paleontology, and the modern synthesis. In *The Evolutionary Synthesis*, ed. E. Mayr and W. Provine, pp. 153–72. Cambridge, MA: Harvard University Press.

  (1985). A clock of evolution. *Natural History*, **94**, 12–25.

Haldane, J. B. S. (1964). A defense of beanbag genetics. *Perspectives in Biology and Medicine*, **7**, 343–59.

Kao, F.-T. (1985). Human genome structure. *International Review of Cytology*, **96**, 51–201.

Korenberg, J. R. and Rykowski, M. (1988). Human genome organization: Alu, Lines, and the molecular structure of metaphase chromosome bands. *Cell*, **53**, 391–400.

Krystal, M., D'Eustachio, P., Ruddle, F. and Arnheim, N. (1981). Human nucleolus organizers on non-homologous chromosomes can share the same ribosomal gene variants. *Proceedings of the National Academy of Sciences USA*, **78**, 5744–8.

Lehrman, M. A., Goldstein, G., Russell, D. and Brown, M. S. (1987). Duplication of seven exons in LDL receptor gene caused by Alu-Alu recombination in a subject with familial hypercholesterolemia. *Cell*, **48**, 827–35.

Manuelidis, L. and Wu, J. C. (1978). Homology between human and simian repeated DNA. *Nature*, **276**, 92–4.

Marks, J. (1987). Social and ecological aspects of primate cytogenetics. In *The Evolution of Human Behavior: Primate Models*, ed. W. Kinzey, pp. 139–50. Albany, NY: SUNY Press.

(1989). Molecular micro- and macro-evolution in the primate alpha-globin gene family. *American Journal of Human Biology*, **1**, 555–66.

Mayr, E. (1959). Where are we? *Cold Spring Harbor Symposia on Quantitative Biology*, **24**, 1–12.

McClintock, B. (1977). The significance of responses of the genome to challenge. *Science*, **226**, 792–801.

Michod, R. (1981). Positive heuristics in evolutionary biology. *British Journal of Philosophy of Science*, **32**, 1–36.

Miyamoto, M., Slightom, J. and Goodman, M. (1987). Phylogenetic relations of humans and African apes from DNA sequences in the pseudo-eta globin gene region. *Science*, **238**, 369–73.

Orgel, L. C. and Crick, R. H. C. (1980). Selfish DNA: The ultimate parasite. *Nature*, **284**, 604–7.

Orkin, S. (1987). The pitfalls of sequencing: revised sequence of the X-CGD gene. *Trends in Genetics*, **3**, 207.

Paulson, K. E. and Schmid, C. W. (1986). Transcriptional inactivity of Alu repeats in HeLa cells. *Nucleic Acids Research*, **14**, 6145–58.

Penny, D. (1989). What, if anything, is *Prochloron*? *Nature*, **337**, 304–5.

Provine, W. B. (1986). *Sewall Wright and Evolutionary Biology*. Chicago, IL: University of Chicago Press.

Quattrochi, L., Pendurthi, U., Okino, S., Potenza, C. and Tukey, R. H. (1986). Human cytochrome P-450 4 mRNA and gene: Part of a multi-gene family that contains *Alu* sequences in its mRNA. *Proceedings of the National Academy of Sciences USA*, **83**, 6731–5.

Rose, M. R. and Doolittle, W. F. (1983). Molecular biological mechanisms of speciation. *Science*, **220**, 157–62.

Rosenberg, H., Singer, M. and Rosenberg, M. (1978). Highly reiterated sequences of SIMIANSIMIANSIMIANSIMIANSIMIAN. *Science*, **200**, 394–402.

Rouyer, F., Simmier, M.-C., Page, D. and Weissenback, J. (1987). A sex chromosome rearrangement in a human XX male caused by Alu-Alu recombination. *Cell*, **51**, 417–25.

Schmid, C. W., Fox, G., Dowds, B., Lowensteiner, L., Paulson, E., Shen, C.-K. J. and Leinwand, L. (1983). Families of repeated human DNA sequences and their arrangements. In *Perspectives on Genes and the Molecular Biology of Cancer*, ed. D. L. Robberson and G. F. Saunders, pp. 35–41. New York: Raven Press.

Schmid, C. W. and Jelinek, W. (1982). The Alu family of dispersed repetitive repeats. *Science*, **216**, 1065–70.

Schmid, C. W. and Marks, J. (1990). DNA hybridization as a guide to phylogeny: Physical and chemical limits. *Journal of Molecular Evolution*, **30**, 237–46.

Schmid, C. W. and Shen, C.-K. J. (1985). The evolution of interspersed repetitive DNA sequences in mammals and other vertebrates. In *Molecular Evolutionary Genetics*, ed. R. J. MacIntyre, pp. 323–58. New York: Plenum Press.

Schughart, K., Kappen, C. and Ruddle, F. H. (1989). Duplication of large genomic regions during the evolution of vertebrate homeobox genes. *Proceedings of the National Academy of Sciences USA*, **86**, 7067–71.

Snyder, M. and Doolittle, W. F. (1988). P elements in *Drosophila*: selection at many levels. *Trends in Genetics*, **4**, 147–9.

Sparrow, A. H. and Nauman, A. F. (1976). Evolution of genome size by DNA doubling. *Science*, **192**, 524–9.

Stebbins, G. L. (1966). Chromosomal variation and evolution. *Science*, **152**, 1463–9.

Stoppa-Lyonnet, D., Carter, P., Meo, T. and Tosi, M. (1990). Clusters of intragenic *Alu* repeats predispose the human C1 inhibitor locus to deleterious rearrangements. *Proceedings of the National Academy of Sciences USA*, **87**, 1551–5.

Stout, J. T. and Carey, C. T. (1988). The Lesch–Nyhan syndrome: clinical, molecular and genetic aspects. *Trends in Genetics*, **4**, 175–8.

Syvanen, M. (1984). The evolutionary implications of mobile genetic elements. *Annual Review of Genetics*, **18**, 271–93.

Temin, H. and Engels, W. (1984). Movable genetic elements and evolution. In *Evolutionary Theory: Paths into the Future*, ed. J. W. Pollard, pp. 173–201. New York: John Wiley.

Tolan, D. R., Niclas, J., Bruce, B. D. and Lebo, R. V. (1987). Evolutionary implications of the human aldolase-A, -B, -C, and -pseudogene chromosome locations. *American Journal of Human Genetics*, **41**, 907–24.

Vrba, E. and Eldredge, N. (1984). Individuals, hierarchies and processes: Towards a more complete evolutionary theory. *Paleobiology*, **10**, 146–71.

Waye, J. S. and Willard, H. F. (1989). Concerted evolution of alpha satellite DNA: Evidence for species specificity and a general lack of sequence conservation among alphoid sequences of higher primates. *Chromosoma*, **98**, 273–9.

Weiner, A., Deininger, P. L. and Efstradiatis, A. (1986). Non-viral retroposons: Genes, pseudogenes, and transposable elements generated by the reverse flow of genetic information. *Annual Review of Biochemistry*, **55**, 631–61.

Wevrick, R. and Willard, H. F. (1989). Long-range organization of tandem arrays of alpha satellite DNA at the centromeres of human chromosomes: High-frequency array-length polymorphism and meiotic stability. *Proceedings of the National Academy of Sciences USA*, **86**, 9394–8.

White, M. J. D. (1969). Chromosomal rearrangements and speciation in animals. *Annual Review of Genetics*, **3**, 75–98.

   (1978). Chain processes in chromosomal speciation. *Systematic Zoology*, **27**, 285–98.

Williams, S. M., De Bry, R. W. and Feder, J. L. (1988). A commentary on the use of ribosomal DNA in systematic studies. *Systematic Zoology, 37*, 60–2.

Wright, S. (1931). Evolution in Mendelian populations. *Genetics*, **16**, 97–159.

   (1932). General, group and special size factors. *Genetics*, **17**, 603–19.

Index

atherosclerosis 56–75

baboon 27, 40, 41, 56–75, 85

cDNA, *see* complementary DNA
candidate genes
   atherosclerosis 60, 68
chimpanzee 26, 42–44
complementary DNA 67–8

DNA
   extraction 60–1, 187–8
   fingerprinting 23, 83
      colony management 91
      paternity testing 83–6
   mitochondrial 179–233
      isolation 186–7
      mammals, non-primate 182
      primate, non-human 182–6, 197–205
   nuclear 19–55
      mammals, non-primate 85, 90
      primate, non-human 26, 27, 40, 41,
         42–4, 56–75, 84, 85, 86
   see also individual primate name
   sequences 32
   sequencing 1, 191, 247–8

gene conversion 106–8, 138
genetic drift 122, 125
genome 234–55
   human 20
   mitochondrial 180–2
gibbon 27
globin
   alpha 30, 37, 41, 103–78, 238–9
   beta 9, 26, 34–5, 37, 39, 41, 103–78,
      238
   deletions 123–7
   delta 26, 30, 103–78
   gamma 30, 103–78
   zeta 239
gorilla 24, 42–4

HDL, *see* lipoproteins
HOX, *see* homeobox
haemoglobinopathies 121–60
haplotype 9, 32–4, 109
   globin 36–9, 103–78

homeobox 41–5
hybridization
   DNA–DNA 43, 245–6
isoschizomer 3

LDL, *see* lipoproteins
lemur 90
library
   genomic 10, 15, 66
linkage disequilibrium 109
lipoproteins 56–75

mtDNA, *see* DNA, mitochondrial
macaque 84, 86
map
   linkage 12
minisatellite 13–16, 22–4, 34, 76–102,
   108, 127
molecular drive 243, 244
mutation
   CpG dimer 7, 11
   depletion 106–8
   duplication 29, 31
   globin 104–8
   insertion/deletion 22, 31
   point 21, 22, 105
   rearrangement 24–9

natural selection 39–41, 119, 128–60

orangutan 42–4

P elements 243
PCR, *see* polymerase chain reaction
PIC value, *see* polymorphism information
   content
PKU, *see* phenylketonuria
paralog 249
phenylketonuria 40
phylogeny 33, 41–4, 183, 196–7
polymerase chain reaction 1, 70–1, 94–5,
   193
polymorphism 34–9
   information content 9, 10
polyploidy 243
population structure 44, 103–78, 202–3
pseudogene 26

257

Printed in the United States
By Bookmasters